AF381398

Array Signal Processing

S. Uṇṇikrishṇa Pillai

Array Signal Processing

C.S. Burrus
Consulting Editor

Springer-Verlag
New York Berlin Heidelberg
London Paris Tokyo

S. Uṇṇikrishṇa Pillai
Department of Electrical Engineering and
  Computer Science
Polytechnic University
Brooklyn, New York 11201
USA

Consulting Editor
Signal Processing and Digital Filtering

C.S. Burrus
Professor and Chairman
Department of Electrical and
  Computer Engineering
Rice University
Houston, TX 77251-1892
USA

Library of Congress Cataloging-in-Publication Data
Pillai, S. U.
    Array signal processing / S. Uṇṇikrishṇa Pillai ; C.S. Burrus,
  consulting editor.
      p.   cm.
    Includes index.

    1. Signal processing.   I. Burrus, C.S.   II. Title.
  TK5102.5.P56 1989   88-37026
  621.38′043—dc19

Printed on acid-free paper

Camera-ready copy prepared by the author using eroff.

9 8 7 6 5 4 3 2 1
ISBN-13: 978-1-4612-8186-3   e-ISBN-13: 978-1-4612-3632-0
DOI: 10.1007/978-1-4612-3632-0

To

Professor Dante C. Youla
and the illustrious forefathers
Pāṇini and Bādarāyaṇa

Preface

This book is intended as an introduction to array signal processing, where the principal objectives are to make use of the available multiple sensor information in an efficient manner to detect and possibly estimate the signals and their parameters present in the scene. The advantages of using an array in place of a single receiver have extended its applicability into many fields including radar, sonar, communications, astronomy, seismology and ultrasonics. The primary emphasis here is to focus on the detection problem and the estimation problem from a signal processing viewpoint. Most of the contents are derived from readily available sources in the literature, although a certain amount of original material has been included.

This book can be used both as a graduate textbook and as a reference book for engineers and researchers. The material presented here can be readily understood by readers having a background in basic probability theory and stochastic processes. A preliminary course in detection and estimation theory, though not essential, may make the reading easy. In fact this book can be used in a one semester course following probability theory and stochastic processes. Concepts are explained and illustrated in detail along with important mathematical techniques. Since it is much easier to skip steps than to reconstruct them, complete proofs of all the major results are included and the book is essentially self contained. Problems at the end of each chapter have been chosen to extend the material presented in the book.

I wish to take this opportunity to thank my colleagues at Polytechnic University, especially Professor Dante C. Youla for his criticisms, encouragement and many useful comments through various stages of this manuscript. Working with Professor Youla on research problems has been a rewarding and quite satisfying experience. My students Youngjik Lee and Byung Ho Kwon have gone through the entire manuscript and worked out several computations in chapter 3. Y. Lee also helped me with major parts of the word processing work. Their willingness and enthusiasm to help are gratefully acknowledged.

Special thanks to Michael Rosse for editing the typed manuscript and Leonard Shaw, Fred Haber, Saleem A. Kassam, Sid Burrus, Julia Abrahams, Rabinder N. Madan and my friends Brig Elliott and Joan McEntee Mariani for their kind encouragement. I would like to acknowledge the research support received from the Office of Naval Research, which eventually prompted me to write this book.

Finally a word about the Sanskrit scholars Pānini and Bādarāyana mentioned in the dedication. The great grammarian Pānini (ca. 500 B.C.), and Bādarāyana (ca. 350 B.C.) to whom the Brhadaranyaka Upanishad and Brahma Sūtra are attributed, are my all time heros. Of course, Professor Youla is contemporary and continues to be a source of good ideas.

Brooklyn, 1988   S. Unnikrishna Pillai

Contents

Chapter 3   Performance Analysis

Chapter 4   Estimation of Multiple Signals

Array Signal Processing

Chapter 1
Introduction

1.1 Introduction

Sensor arrays have been in use for several decades in many practical signal processing applications. Such an array consists of a set of sensors that are spatially distributed at known locations with reference to a common reference point. These sensors collect signals from sources in their field of view. Depending on the sensor characteristics and the path of propagation, the source waveforms undergo deterministic and/or random modifications. The sensor outputs are composed of these source components and additive noise such as measurement and thermal noise.

In active sensing situations such as radar and sonar, a known waveform of finite duration is generated which in turn propagates through a medium and is reflected by some target back to the point of origin. The transmitted signal is usually modified both in amplitude and phase by the target characteristics, which by themselves might be changing with time and its position in space. These disturbances give rise to a random return signal. In the passive context the signal received at the array is self-generated by the target, such as propeller or engine noise from submarines in the case of sonar. Once again the signals are random in nature. In addition to these target-generated direct signals, there may also be spurious returns such as clutter in the case of radar. Moreover, signals from a target can undergo reflection, creating multiple returns that are delayed, amplitude-weighted replicas of the direct signal to the array. These, as well as intentional jamming signals, generate coherent interference. For example, in radar, multipath returns give rise to secondary signals that are completely coherent with the original signal. Similar phenomena occur in sonar, where reverberations from seabed of an incoming signal create multiple echos. In a satellite communication scenario, a smart jammer may artfully create such a signal to neutralize the incoming desired signal. In general, the signals may be uncorrelated, partially correlated or completely coherent with each other. Similarly, the additive noise contained at the sensor outputs may also be uncorrelated or correlated with each other and may have equal/unequal noise variances. Since the natural causes responsible for signals and noise are often unrelated, it is customary to assume that the signals and noise are uncorrelated with each other.

Signals that can be adequately characterized by a single frequency are known as narrowband signals. In contrast to this, signals that occupy a significant frequency band constitute broadband or wideband sources. Physically, the signals may have originated far away from the array, or their point of origin can be quite close to the array. The former case is referred to as the far-field situation, and, by virtue of the distance between the sources and the array, the information carrying wavefronts associated with these far-field sources at the array may be assumed to be planar. In case of narrowband signals, if these plane waves advance through a nondispersive medium that only introduces propagation delays, the output of any other array element can be represented by a time-advanced or time-delayed version of the signal at the reference element.

The practical problems of interest in array signal processing are extracting the desired parameters such as the directions of arrival, power levels and crosscorrelations of the signals present in the scene from the available information including the measured data. Often one may also be specifically interested in the actual signal of one of these sources, and in that case it is necessary to estimate the actual waveform associated with the desired signal by improving the overall reception in an environment having several sources. To achieve this, ideally it should be possible to suppress the undesired signals and enhance the desired signal. The desired signal may correspond to a friendly satellite signal in presence of hostile jammers which may have time varying characteristics. This can happen because of physical motion or deliberate on-off jamming strategies of the opponent. In this case, quick adaptive learning capabilities of the changing scene are required to maintain an acceptable level of the desired signal characteristics at the receiver. At times, the desired signal structure might be partially known, and the objective in that case is to detect its presence in the available noisy data. This situation is often encountered in sonar where the data is analyzed at the receiver to detect the presence of the signature of a specific class of submarine. Though the signal structure is known, it may still contain unknown parameters such as angle of arrival or random phase.

All these problems fall into one of the two categories: detection or estimation of signals and their parameters from multichannel information. For the signal and noise models discussed above, they form part of our study here.

1.2 Organization of the Book

Chapter 2 introduces the array concept and examines the signal and noise model it is expected to receive in detail. The advantages in using an array are illustrated for a single source scene, and the array structure and its impact on general performance is discussed.

The problem of detecting multiple signals is examined in great detail beginning with traditional techniques that estimate the signal parameters such as their arrival angles and power levels from direction dependent array output power measurements. This is followed by eigenstructure-based high resolution methods that exploit certain structural properties of the array output covariance matrix to achieve the same goal. The modifications of these methods to include coherent source scenes, and related schemes are also examined. The discussion in chapter 2 assumes exact knowledge of the required second order information, i.e., the array output covariance matrix is assumed to be completely known.

Chapter 3 analyzes the performance of the above mentioned techniques, when in the absence of the ensemble averages, these covariances are estimated directly from the available noisy data. Beginning with the maximum likelihood estimate of the covariance matrix in the case of zero mean, complex circular Gaussian data vectors, a detailed asymptotic analysis of the eigenvalues and a certain set of eigenvectors of this estimated covariance matrix is presented. This in turn is used to derive the mean and variance and in some cases the probability distribution function of the angle-of-arrival estimators discussed in chapter 2. In addition, using this analysis, asymptotic results for resolving two closely spaced sources are derived in terms of signal-to-noise ratio, number of array elements and the signal correlation. Comparisons are also presented to illustrate their performances relative to each other.

The signal acquisition problem discussed in chapter 4 first formulates the Wiener solution, which is optimum under the minimization of the mean square error criterion. This is followed by other optimality criteria such as maximization of signal-to-noise ratio and maximum likelihood performance measure. The realization of the Wiener solution through adaptive recursive procedures is outlined, and analysis of the least mean square technique is presented with details regarding its convergence properties. The chapter concludes with direct implementation techniques for realizing the Wiener solution.

1.3 Notations and Preliminaries

Throughout this book scalar quantities are denoted by regular lower or upper case letters. Lower and upper case bold type faces are used for vectors and matrices respectively. Thus a (or A), $\mathbf{a}$ and $\mathbf{A}$ stand for scalar, vector and matrix in that order. Similarly $\mathbf{A}^*$, $\mathbf{A}^T$, $\mathbf{A}^\dagger$, $tr(\mathbf{A})$ and $det(\mathbf{A}) = |\mathbf{A}|$ represent the complex conjugate, transpose, complex conjugate transpose, trace and determinant of $\mathbf{A}$ respectively. The symbol $diag[\lambda_1, \lambda_2, \cdots, \lambda_M]$ is used sometimes to represent a diagonal matrix with diagonal entries $\lambda_1, \lambda_2, \cdots, \lambda_M$. For two square matrices $\mathbf{A}$, $\mathbf{B}$ of same size, $(\mathbf{AB})^\dagger = \mathbf{B}^\dagger \mathbf{A}^\dagger$, $tr(\mathbf{AB}) = tr(\mathbf{BA})$ and $|\mathbf{AB}| = |\mathbf{A}||\mathbf{B}|$.

A square matrix $\mathbf{A}$ of size $M \times M$ is said to be hermitian if $\mathbf{A} = \mathbf{A}^\dagger$, i.e., $a_{ij} = a_{ji}^*$ for all i, j, where a_{ij} represents the $(i,j)^{\text{th}}$ element of $\mathbf{A}$. It is said to be nonnegative definite if for any $M \times 1$ vector $\mathbf{x}$, $\mathbf{x}^\dagger \mathbf{A} \mathbf{x} \geq 0$. When strict inequality holds, i.e., $\mathbf{x}^\dagger \mathbf{A} \mathbf{x} > 0$ for $\mathbf{x} \neq 0$, $\mathbf{A}$ is said to be positive definite.

Let λ_i denote an eigenvalue of $\mathbf{A}$. Then, there exists an eigenvector $\mathbf{u}_i \neq 0$, such that $\mathbf{A}\mathbf{u}_i = \lambda_i \mathbf{u}_i$. The eigenvectors are in general complex and for positive definite matrices, they can be made unique by normalization together with a constraint of the form $u_{ii} \geq 0$. The eigenvalues of a hermitian matrix are real. For, $\mathbf{u}_i^\dagger \mathbf{A} \mathbf{u}_i = \lambda_i \mathbf{u}_i^\dagger \mathbf{u}_i = \lambda_i$. Also $\mathbf{u}_i^\dagger \mathbf{A} \mathbf{u}_i = (\mathbf{A}\mathbf{u}_i)^\dagger \mathbf{u}_i = \lambda_i^* \mathbf{u}_i^\dagger \mathbf{u}_i = \lambda_i^*$. Thus $\lambda_i^* = \lambda_i$ or λ_i is real. If $\mathbf{A}$ is also positive definite then it follows that its eigenvalues are all positive. Moreover, eigenvectors associated with distinct eigenvalues of a hermitian matrix are orthogonal. This follows by noticing that, if λ_i, λ_j and $\mathbf{u}_i$, $\mathbf{u}_j$ are any two such pairs, then $\mathbf{u}_i^\dagger \mathbf{A} \mathbf{u}_j = \mathbf{u}_i^\dagger \lambda_j \mathbf{u}_j = \lambda_j \mathbf{u}_i^\dagger \mathbf{u}_j$. However, we also have $\mathbf{u}_i^\dagger \mathbf{A} \mathbf{u}_j = (\mathbf{A}\mathbf{u}_i)^\dagger \mathbf{u}_j = (\lambda_i \mathbf{u}_i)^\dagger \mathbf{u}_j =$

$\lambda_i \mathbf{u}_i^\dagger \mathbf{u}_j$. Thus, $\lambda_j \mathbf{u}_i^\dagger \mathbf{u}_j = \lambda_i \mathbf{u}_i^\dagger \mathbf{u}_j$ or equivalently $\mathbf{u}_i^\dagger \mathbf{u}_j = 0$ provided $\lambda_i \neq \lambda_j$. Further, this implies that if an eigenvalue repeats (say L times), then the associated eigenvectors span an L-dimensional subspace that is orthogonal to the one spanned by the remaining set of eigenvectors. As a result, it is always possible to choose a new set of L orthonormal vectors from the above L-dimensional subspace to act as an eigenvector set for the above repeating eigenvalue. Thus, for an $M \times M$ hermitian matrix $\mathbf{A}$, if $\lambda_1, \lambda_2, \cdots, \lambda_M$ and $\mathbf{u}_1, \mathbf{u}_2, \cdots, \mathbf{u}_M$ represent its eigenvalues and an orthonormal set of eigenvectors, then $\mathbf{A}\mathbf{u}_i = \mathbf{u}_i \lambda_i$, $i = 1, 2, \cdots, M$, or in a compact form $\mathbf{AU} = \mathbf{U\Lambda}$, where $\mathbf{U} = [\mathbf{u}_1, \mathbf{u}_2, \cdots, \mathbf{u}_M]$ and $\mathbf{\Lambda} = diag[\lambda_1, \lambda_2, \cdots, \lambda_M]$. Clearly, $\mathbf{U}\mathbf{U}^\dagger = \mathbf{U}^\dagger \mathbf{U} = \mathbf{I}$, i.e., $\mathbf{U}$ is a unitary matrix, and consequently $\mathbf{AU} = \mathbf{U\Lambda}$ gives $\mathbf{A} = \mathbf{U\Lambda U}^\dagger$. Thus, any hermitian matrix can be diagonalized by a unitary matrix whose columns represent a complete set of its normalized eigenvectors. Moreover $|\mathbf{A}| = |\mathbf{U}| |\mathbf{\Lambda}| |\mathbf{U}^\dagger| = \lambda_1 \lambda_2 \cdots \lambda_M$ and $tr(\mathbf{A}) = \lambda_1 + \lambda_2 + \cdots + \lambda_M$.

The number of linearly independent rows (or columns) of a matrix $\mathbf{A}$ represents its rank $\rho(\mathbf{A})$. In case of a square matrix, its rank coincides with the total number of nonzero eigenvalues (including repetitions). For any two matrices $\mathbf{A}$ and $\mathbf{B}$ of dimensions $m \times n$ and $n \times r$, the rank of their product satisfies Sylvester's inequality [1] given by $\rho(\mathbf{A}) + \rho(\mathbf{B}) - n \leq \rho(\mathbf{AB}) \leq min[\rho(\mathbf{A}), \rho(\mathbf{B})]$. A matrix of full rank is said to be nonsingular. Any nonnegative definite (hermitian) matrix may be factored into the form $\mathbf{A} = \mathbf{C}^2$ where $\mathbf{C}$ is also hermitian and $\rho(\mathbf{A}) = \rho(\mathbf{C})$. A good knowledge of these results is essential to understanding the rest of this book.

Reference

[1]   F. R. Gantmacher, *The Theory of Matrices*. New York: Chelsea, 1977.

Chapter 2
Detection of Multiple Signals

2.1 Signals and Noise

In this chapter we will discuss the problem of detecting multiple signals using information from multiple sensors. To understand the advantages of using a sensor array over a single element in various aspects of detection and estimation it is necessary to understand the nature of signals and noise the array is desired to receive.

In active sensing situations such as radar and sonar, a known waveform of finite duration is generated which in turn propagates through a medium and is reflected by some target back to the point of origin. The transmitted signal is usually modified both in amplitude and phase by the target characteristics, which by themselves might be changing with time and its position in space. These disturbances give rise to a random return signal. In the passive context the signal received at the array is self-generated by the target, such as propeller or engine noise from submarines in the case of sonar. Once again the signals are random in nature. In addition to these target-generated direct signals, there may also be spurious returns such as clutter in the case of radar and reverberations from the ocean surface/bed or sea layers in the case of sonar. Moreover, signals from a target can undergo reflection, creating multipath returns that are delayed, amplitude-weighted replicas of the direct signal to the array. These as well as intentional jamming signals can generate coherent interference. In all these cases the signals that arrive at the array can be regarded as random, and at times the physical phenomena responsible for the randomness in the signal make it plausible to assume that the signals are Gaussian (normal) random processes.

Likewise thermal sensor noise and ambient noise are also random in nature. These additive components at the sensor outputs usually represent the totality of several small independent and identical sources, and application of the central limit theorem permits one to model the resulting noise as a Gaussian and (usually) stationary process. Needless to say, in such situations the totality of signal and noise process can be completely specified by their first and second order moment characterization.

At any instant every signal has an amplitude and phase (with respect to a reference) component. At times, in addition to this in-phase component, it becomes necessary to generate its quadrature part for optimum processing [1]. Instead of carrying out all computations separately in terms of their in-phase and quadrature parts, it is often convenient and advantageous to represent them as the real and imaginary parts of a complex signal. The actual signals that appear in any physical system are real and in this representation they appear as the real part of the equivalent complex signal [2, 3]. Thus at a reference element if

$$u_r(t) = \sqrt{P} \, \cos(\omega_0 t + \phi(t) + \theta)$$

is the actual phase modulated carrier signal with a random phase factor θ, its complex representation is

$$\bar{u}(t) = \sqrt{P} \, e^{(\omega_0 t + \phi(t) + \theta)} . \tag{2.1}$$

Clearly,

$$u_r(t) = Re\,[\bar{u}(t)]$$

where $Re\,[\cdot]$ stands for the real part of $[\cdot]$ and

$$u(t) = \sqrt{P} \, e^{(\phi(t) + \theta)} \tag{2.2}$$

is known as the baseband reduced complex envelope of the real signal $u_r(t)$. Thus only the information carrying component, and not the carrier frequency of the modulated signal, appear in this description. In physical terms, if the signals have originated far away from the array then they can be modeled as information carrying uniform plane waves as above. If these plane waves advance through a non-dispersive medium that only introduces propagation delays, the output of any other array element can be represented by a time-advanced or time-delayed version of the above complex envelope at the reference element. Once again this avoids the carrier frequency description.

Signals such as above that can be adequately characterized by a single carrier frequency are known as narrowband sources. In contrast to this, signals that occupy a significant spectral band constitute broadband (wideband) sources. Further all of these signals can be uncorrelated, correlated or coherent with each other. In particular, for two jointly stationary signals $u_i(t)$, $u_j(t)$, let ρ_{ij} represent their correlation coefficient [4]. By definition,

$$\rho_{ij} = \frac{E[u_i(t)u_j^*(t)]}{\sqrt{E[\,|u_i(t)|^2]E[\,|u_j(t)|^2]}} \qquad (2.3)$$

and $|\rho_{ij}| \leq 1$ as follows from the Schwarz inequality. Thus,

$$\rho_{ij} = 0 \quad \rightarrow \quad u_i(t), u_j(t) \text{ are uncorrelated}$$

$$0 < |\rho_{ij}| < 1 \quad \rightarrow \quad u_i(t), u_j(t) \text{ are correlated} \qquad (2.4)$$

$$|\rho_{ij}| = 1 \quad \rightarrow \quad u_i(t), u_j(t) \text{ are coherent}.$$

It is not difficult to show that under coherent conditions

$$u_j(t) = \alpha u_i(t)$$

where α is a nonrandom complex constant. In practice the additive noise components are uncorrelated with the signal part. However, among themselves the interelement noises may be correlated or uncorrelated with each other. One standard assumption is to consider them as uncorrelated (independent in case of Gaussian) and identical processes, i.e., if $n_i(t)$, $n_j(t)$ represent i^{th} and j^{th} sensor noise, then

$$E[n_i(t)n_j^*(t)] = \sigma^2 \delta_{ij} \qquad (2.5)$$

where σ^2 represents the noise power common to all sensors.

It was remarked earlier that when an advancing plane wave passes through a non-dispersive medium, the signal output at any sensor element immersed in that medium can be represented as a time-delayed/advanced version of its complex envelope at a reference element. To see this, let $\bar{u}(t)$ in (2.1) denote the complex signal representing a modulated plane wave at the reference element in Fig.

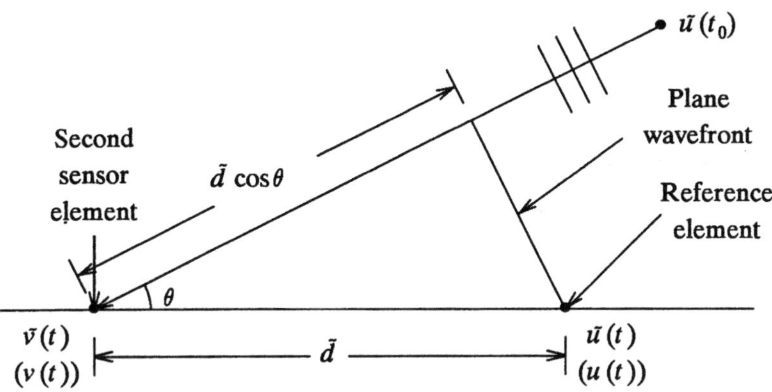

Fig. 2.1 Pair of identical sensor elements. $u(t)$, $v(t)$ represent the corresponding complex envelope signals associated with $\tilde{u}(t)$ and $\tilde{v}(t)$.

2.1. The normal to the plane wavefront makes an angle θ with the line joining the sensors in the linear array. Further let $\tilde{d}$ denote the distance of the second sensor from the reference point in absolute units. The output $\tilde{v}(t)$ at the second sensor is delayed by the time required for the plane wave to propagate through $\tilde{d}\cos\theta$ and if c represents the velocity of propagation, then this time delay τ is given by

$$\tau = \frac{\tilde{d}\cos\theta}{c}. \tag{2.6}$$

Thus

$$\tilde{v}(t) = \tilde{u}(t-\tau). \tag{2.7}$$

If the carrier frequency is fairly large compared to the bandwidth of the modulating signal, then the modulating signal can be treated as quasi-static during time intervals of order τ and in that case (2.7) reduces to

$$\tilde{v}(t) = \tilde{u}(t)e^{\frac{-j\,\omega_0\tilde{d}\cos\theta}{c}} = \tilde{u}(t)e^{-j\,2\pi\frac{\tilde{d}\cos\theta}{\lambda}} \tag{2.8}$$

where λ is the associated carrier wavelength. In terms of the baseband reduced complex envelope representation, the reference signal $u(t)$ is given by (2.2) and from (2.8) the second sensor output is

$$v(t) = u(t)e^{-j\,2\pi\tilde{d}\cos\theta/\lambda}. \tag{2.9}$$

Often it is convenient to express the interelement spacing in terms of normalized dimensionless units. For reasons that will be explained soon, it is convenient to normalize all distances with respect to half wavelength. Let d represent the normalized distance between the reference element and the second sensor. Then $d = \tilde{d}/(\lambda/2)$ and (2.9) becomes

$$v(t) = u(t)e^{-j\,\pi d\cos\theta}. \tag{2.10}$$

Not surprisingly, for narrowband signals, the time delay appears as a pure phase delay of the reference signal. Moreover, this phase delay depends only on the spacing between the sensors in question and the angle of arrival of the plane wave, and is independent of the time variable. However such is not the case in a broadband situation and if the complex envelope $u(t)$ at the reference element represents a broadband signal, then the corresponding output $v(t)$ at the second sensor in Fig. 2.1 can be written as

$$v(t) = u(t-\tau) \tag{2.11}$$

with τ as in (2.6). With $U(f)$ and $V(f)$ representing the Fourier transforms of $u(t)$ and $v(t)$ respectively, (2.11) reduces to

$$V(f) = U(f)e^{-j\,2\pi f\,\tau} = U(f)e^{-j\,\pi f\,d\cos\theta/f_0}. \tag{2.12}$$

Notice that (2.12) is structurally identical to the narrowband situation represented in (2.10) and hence conceptually, at least, techniques devised for narrowband cases can be applied to the broadband signals in the frequency domain. However the phase delay in (2.12) is frequency sensitive (function of the free variable f), and this is to be contrasted with (2.10) where it is independent of the free variable t. This important difference should be taken into consideration while dealing with broadband information.

Advantages in Using an Array

The possibility of modifying the array outputs to enhance the desired signal reception and simultaneously suppress the undesired ones can be illustrated by considering a single source situation as in

Fig. 2.1 in presence of M identical sensors. Let $d_1, d_2, \cdots, d_M$ represent the normalized distances of these sensors with respect to a reference point and $u(t)$ the complex envelope of the signal at that point. Further let $n_1(t), n_2(t), \cdots, n_M(t)$ represent the respective noise components that are assumed to be independent and identical as in (2.5). With $x_i(t)$ representing the complex envelope of the total received signal at the ith sensor, and using (2.10) it is easy to see that

$$x_i(t) = u(t)e^{-j\pi d_1 \cos\theta} + n_i(t) \tag{2.13}$$

and the input signal-to-noise-ratio (SNR) is

$$(SNR)_i = \frac{E[\,|u(t)|^2\,]}{E[\,|n_i(t)|^2\,]} = \frac{P}{\sigma^2}$$

where $P \triangleq E[\,|u(t)|^2\,]$ represents the signal power. From (2.13) the signal components can be coherently combined if the array outputs are phase shifted by $e^{j\pi d_1 \cos\theta}$; $i = 1, 2, \cdots, M$ and the resulting signals are added up. This gives the output signal $y(t)$ to be

$$y(t) = \sum_{i=1}^{M} x_i(t)e^{j\pi d_1 \cos\theta} = M u(t) + \sum_{i=1}^{M} n_i(t)e^{j\pi d_1 \cos\theta} = M u(t) + n(t).$$

The output SNR in this case is given by

$$(SNR)_o = \frac{E[\,|M u(t)|^2\,]}{E[\,|n(t)|^2\,]} = \frac{M^2 P}{|\sum_i \sum_j E[n_i(t)n_j^*(t)]|}$$

$$= \frac{M^2 P}{M \sigma^2} = M (SNR)_i . \tag{2.14}$$

Thus a simple phase shifting and adding operation among the sensor outputs results in an improvement in the signal-to-noise ratio by a factor equal to the number of sensors. Physically, through appropriate phase delays the desired signal has been coherently combined (in voltage), with noise adding up only incoherently (in power). This results in a gain factor for the overall output signal compared to the noise.

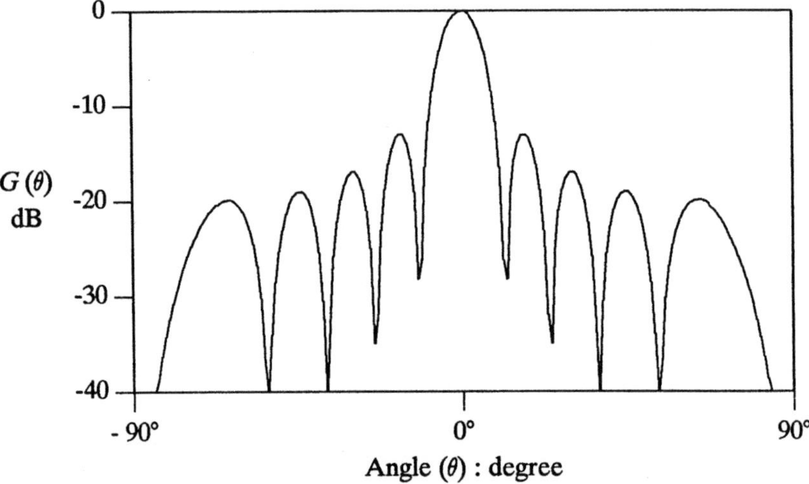

Fig. 2.2 Directional gain pattern for a ten-element uniform array.

Alternatively, the array has been "steered" to look along the direction θ. This process is also known as beamforming. The directional pattern in a plane containing the array may therefore be found from the array factor

$$F(\theta) = \sum_{i=1}^{M} e^{j\pi d_i \cos\theta}$$

and the normalized directional gain pattern

$$G(\theta) = |\frac{1}{M} F(\theta)|^2. \qquad (2.15)$$

For a uniformly placed M element array ($d_i = (i-1)$; $i = 1, 2, \cdots, M$), the above directional pattern has the explicit form

$$G(\theta) = \left(\frac{\sin(\pi M \cos\theta/2)}{M \sin(\pi \cos\theta/2)}\right)^2 \qquad (2.16)$$

and this is plotted in Fig. 2.2 for a ten element array steered along θ. The mainlobe width (beamwidth) is $2\cos^{-1}(2/M)$ and it decreases with increase in the number of sensor elements. The sidelobes represent the gain pattern for signals present along directions other than the look direction while the array is steered along θ and an important question in array design is where to place the array elements for uniformly low sidelobes [5, 6]. Moreover in this setup there always exist $(M\text{-}1)$ null points $(G(\theta) = 0)$ in the field of view (see Fig. 2.2), and in a static situation the array output weights or the interelement distances can be selected to create nulls along the directions of arrival of the undesired sources [6].

The effect of the normalization parameter $\lambda/2$ in the interelement spacing can be best understood in terms of the directivity pattern $G(\theta)$. Suppose the interelement spacings have been normalized with respect to λ. Then $d/\lambda = 1$ and a simple calculation will show that in the associated directional gain pattern the end sidelobes at $\theta \pm 90°$ have a gain equal to the mainlobe since the signal phases now align exactly and add coherently. This results in grating lobes and to avoid them the interelement spacing can at most be $\lambda/2$.

Suppose in addition, for this linear array a constant phase factor of $(i-1)\Delta_0$ is inserted in the ith element of the array for $i = 1, 2, \cdots,$ M. The insertion of this sequence of phase shifts has the effect of shifting the mainlobe by

$$\theta_0 = \cos^{-1}(\Delta_0/\pi),$$

and the overall directional pattern has been steered to this new direction. This effect can be easily incorporated into (2.16) by replacing $\cos\theta$ with $(\cos\theta - \cos\theta_0)$.

So far we have considered only a single source case and the situation is considerably more complicated in a multiple source scene. In that case the parameters of interest include the total number of signals, their respective directions of arrival, associated power levels, and various techniques that have been developed to evaluate these parameters are discussed in detail below.

2.2 Conventional Techniques

Traditionally the array output power is evaluated as a function of the arrival angle under various optimality considerations, and the peaks in the output power distribution are taken to correspond to the true directions of arrival of signals present in the scene. With $x_i(t)$ representing the output at the ith sensor and w_i the corresponding

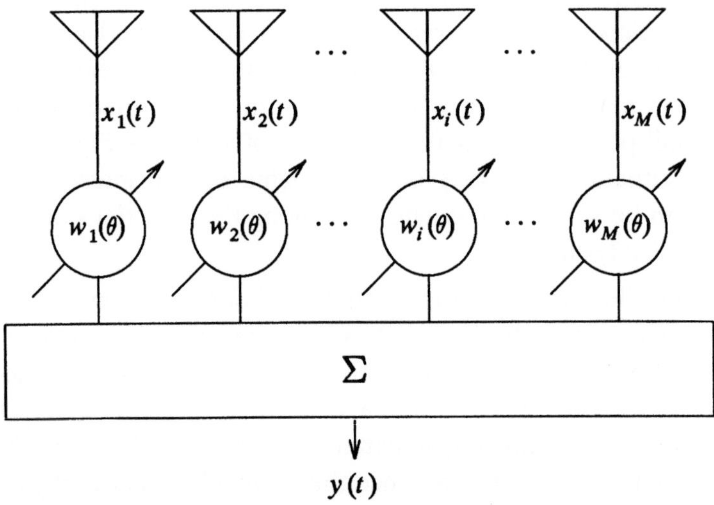

Fig. 2.3 Basic array processing scheme using structured weights.

desired weight factor, using the normalized variable $\omega = \pi \cos\theta$, the array output can be written as (see Fig. 2.3)

$$y(t) = \sum_{i=1}^{M} w_i^*(\omega)x_i(t) = \mathbf{w}^\dagger \mathbf{x}(t) \qquad (2.17)$$

where

$$\mathbf{w} = \left[w_1(\omega), w_2(\omega), \cdots, w_M(\omega) \right]^T \qquad (2.18)$$

stands for the weight vector and

$$\mathbf{x}(t) = \left[x_1(t), x_2(t), \cdots, x_M(t) \right]^T \qquad (2.19)$$

the observed data vector. The average output power $P(\omega)$ is then given by

$$P(\omega) \triangleq E[\,|y(t)|^2] = \mathbf{w}^\dagger E[\mathbf{x}(t)\mathbf{x}^\dagger(t)]\mathbf{w} = \mathbf{w}^\dagger \mathbf{R} \mathbf{w} \qquad (2.20)$$

where

$$\mathbf{R} = E[\mathbf{x}(t)\mathbf{x}^\dagger(t)] \qquad (2.21)$$

represents the $M \times M$ array output covariance matrix. Notice that $\mathbf{R}$ is hermitian ($\mathbf{R} = \mathbf{R}^\dagger$) and always nonnegative definite (i.e., $\mathbf{R} \geq 0$). As shown below, the classical beamforming technique developed by Bartlett and others [7, 8], the minimum variance estimator of Capon [9] and the linear prediction method [10, 11] all fit into this formulation.

2.2.1 Beamformer

As the name implies, the array output weights are chosen to be phase factors required to steer the array along some specific direction θ; i.e.,

$$w_i = \frac{1}{\sqrt{M}} e^{-j\pi d_1 \cos\theta} .$$

For notational convenience define

$$\omega = \pi \cos\theta \qquad (2.22)$$

so that

$$\mathbf{w}_B = \frac{1}{\sqrt{M}} \left[e^{-j d_1 \omega}, e^{-j d_2 \omega}, \cdots, e^{-j d_M \omega} \right]^T \triangleq \mathbf{a}(\omega) . \qquad (2.23)$$

Thus the array output is given by

$$y(t) = \mathbf{w}_B^\dagger \mathbf{x}(t) = \mathbf{a}^\dagger(\omega)\mathbf{x}(t) ,$$

and using (2.20), the output power is

$$P_B(\omega) = E[\,|y(t)|^2\,] = \mathbf{a}^\dagger(\omega)\mathbf{R}\,\mathbf{a}(\omega) . \qquad (2.24)$$

In a single target scene this estimator measures the actual power when steered along the true direction of arrival, resulting in a single peak in that direction. However this is not true even in an uncorrelated multiple source scene and contributions from one source will bias the estimator output along other directions of arrival. This causes the peaks to shift from the true directions of arrival toward each other. It is easily verified that if both sources are located inside the main beam the peaks in fact merge into a single peak, resulting in loss of resolution.

This has led Capon [9] to optimize the weight vector such that while the array is steered along a specific direction, the signal contributions from all other directions are minimized.

2.2.2 Capon's Minimum Variance Estimator

The array output power contains contributions from the desired signal along the look direction as well as the undesired ones along other directions of arrival. To minimize the contributions of this later set, the array output power is minimized here while maintaining the gain along the look direction to be constant. Using (2.20) this is equivalent to the following problem [9]

$$\min_{\mathbf{w}} \ \mathbf{w}^{\dagger}\mathbf{R}\mathbf{w} \ \text{ subject to } \ |\mathbf{w}^{\dagger}\mathbf{a}(\omega)| = 1. \tag{2.25}$$

For positive definite covariance matrices the solution to the weight vector in (2.25) is readily given by

$$\mathbf{w}_C = \frac{\mathbf{R}^{-1}\mathbf{a}(\omega)}{\mathbf{a}^{\dagger}(\omega)\mathbf{R}^{-1}\mathbf{a}(\omega)},$$

and with this weight vector in (2.20) the array output power has the form

$$P_C(\omega) = \frac{1}{\mathbf{a}^{\dagger}(\omega)\mathbf{R}^{-1}\mathbf{a}(\omega)}. \tag{2.26}$$

This procedure is sometimes mistakenly referred to as the "maximum likelihood method" because of the similarity in form of this estimator to that found in the maximum-likelihood estimation of the amplitude of a sine wave of known frequency in Gaussian random noise (see (4.25)).

The objective of any power estimator is to maintain at its output only the power arriving from that specific direction along which the array is being steered. This requires for the array to reject all signals from sources other than those present along the look direction, in turn implying that the weights form a spatial filter with an exact impulse response along the look direction (see Fig. 2.4). Thus while pointing along a specific direction, all arrivals at the array along any other

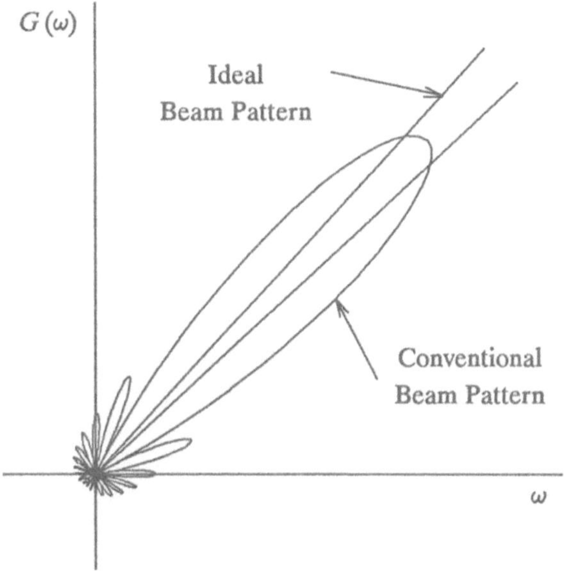

Fig. 2.4 An ideal beamformer.

direction become undesirable, and in this sense they represent direction dependent "signal-like" noise. Consequently, a signal, which is a desired one in the event when the steering direction coincide with its direction of arrival, becomes an undesired signal while the array is being steered along some other direction. The degree of suppression depends on the angular separation between the direction of the signal to be suppressed and the current look direction as well as on the power levels of the signals, the array geometry, and so on. All this information is incorporated into (2.25), which is solved for the weight vector that is used in (2.20) to estimate the actual power.

The estimator in (2.26) has superior resolving power compared to the standard beamformer output in (2.24). This follows from starting with $\mathbf{R}^{1/2}\mathbf{R}^{-1/2} = \mathbf{I}$, which gives

$$\left(\mathbf{a}^{\dagger}(\omega)\,\mathbf{R}^{1/2}\right)\left(\mathbf{R}^{-1/2}\,\mathbf{a}(\omega)\right) = 1$$

and application of Cauchy-Schwarz' inequality readily results in

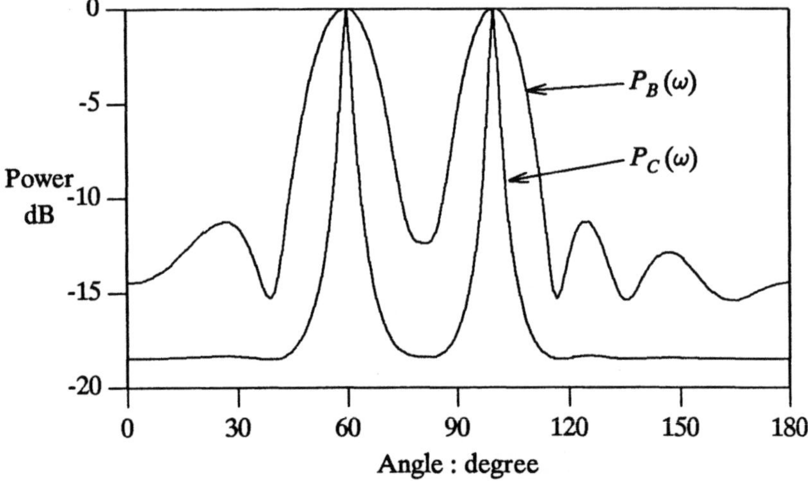

Fig. 2.5 Array output power obtained by the beamformer and the Capon Estimator. Two arrivals from 60° and 100° with a common SNR of 10 dB on a seven-element uniform array. The common maximum value is used to normalize both outputs.

$$\left[\mathbf{a}^{\dagger}(\omega)\,\mathbf{R}\,\mathbf{a}(\omega)\right]\left[\mathbf{a}^{\dagger}(\omega)\,\mathbf{R}^{-1}\mathbf{a}(\omega)\right] \geq 1$$

or

$$P_B(\omega) \geq P_C(\omega). \qquad (2.27)$$

Because of its higher resolving power (Fig. 2.5), the term "high resolution spectral estimator" is also used in reference to this as well as other estimators such as linear prediction method and eigenstructure-based methods. Of these, first we shall explore the linear prediction based method.

2.2.3 Linear Prediction Method

The linear prediction based estimation procedure is commonly used in time series analysis for all pole modeling of the data [12]. It has also been successfully used in array processing [10, 11]. In this case, one of the sensor outputs is predicted as a linear combination of the remaining $(M$-1$)$ sensor outputs at any instant, and the predictor coefficients are selected so as to minimize the mean square error.

Letting $x_n, x_{n-1}, \cdots, x_{n-M+1}$ stand for the M sensor outputs and $\hat{x}_n$ the predictor for x_n, we have

$$\hat{x}_n = -\sum_{i=1}^{M-1} a_i x_{n-i}.$$ (2.28)

This gives the error as

$$\varepsilon_n = x_n - \hat{x}_n = \sum_{i=0}^{M-1} a_i x_{n-i}; \quad a_0 = 1.$$ (2.29)

Minimization of the mean square error $E[\,|\varepsilon_n|^2\,]$ with respect to the unknowns results in the following standard set of linear equations [4]

$$E[\varepsilon_n x_k^*] = \sum_{i=0}^{M-1} a_i E[x_{n-i} x_{n-k}^*] = 0, \quad k = 1, 2, \cdots, M-1$$ (2.30)

and the final mean square error is given by

$$E[\,|\varepsilon_n|^2\,] = E[\varepsilon_n x_n^*] = \sum_{i=0}^{M-1} a_i E[x_{n-i} x_n^*] \triangleq \delta_{M-1}.$$ (2.31)

If the sensor outputs are spatially wide sense stationary, then their cross correlations depend only upon interelement distances, and in that case some interesting conclusions follow. In practice, this occurs when a uniform array receives signals from a set of uncorrelated sources. In that case

$$E[x_{n-i} x_{n-k}^*] = r(k-i) = r^*(i-k)$$ (2.32)

and with this expression in (2.30), (2.31) we have

$$\sum_{i=0}^{M-1} a_i r(k-i) = 0, \quad k = 1, 2, \cdots, M-1,$$

together with

$$\sum_{i=0}^{M-1} a_i r(-i) = \delta_{M-1}.$$

Put together in matrix form this reduces to

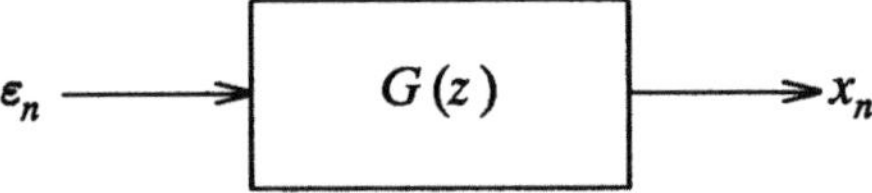

Fig. 2.6 Linear prediction model.

$$
\begin{bmatrix}
r(0) & r(1) & \cdots & r(M\text{-}1) \\
r^*(1) & r(0) & \cdots & r(M\text{-}2) \\
\vdots & \vdots & \ddots & \vdots \\
r^*(M\text{-}1) & r^*(M\text{-}2) & \cdots & r(0)
\end{bmatrix}
\begin{bmatrix}
a_{M\text{-}1} \\
a_{M\text{-}2} \\
\vdots \\
a_1 \\
1
\end{bmatrix}
=
\begin{bmatrix}
0 \\
0 \\
\vdots \\
0 \\
\delta_{M\text{-}1}
\end{bmatrix}. \tag{2.33}
$$

Moreover (2.29) has the system interpretation shown in Fig. 2.6, where

$$
G(z) = \frac{1}{H(z)}
$$

and

$$
H(z) = 1 + a_1 z^{-1} + \cdots + a_{M\text{-}1} z^{-(M\text{-}1)}. \tag{2.34}
$$

Thus x_n can be thought of as the output of a system excited by an uncorrelated noise process (see (2.30)) with average power $\delta_{M\text{-}1}$. Further, if the errors ε_n, ε_{n+k} are also uncorrelated for all n, k, then the input represents a white noise process and x_n an autoregressive process of order $(M\text{-}1)$. From Fig. 2.6, the power spectral density $S_x(\omega)$ of the output process is related to the system transfer function and the input spectral density $S_\varepsilon(\omega) = \delta_{M\text{-}1}$ through the relation

$$
S_x(\omega) = |G(e^{j\omega})|^2 S_\varepsilon(\omega) = \frac{\delta_{M\text{-}1}}{|H(e^{j\omega})|^2}. \tag{2.35}
$$

To simplify this further let $\mathbf{T}_{M\text{-}1}$ denote the hermitian Toeplitz matrix generated by $r(0), r(1), \cdots, r(M\text{-}1)$ and $\Delta_{M\text{-}1}$ its determinant. Then

$$T_{M-1} = \begin{bmatrix} r(0) & r(1) & \cdots & r(M-1) \\ r^*(1) & r(0) & \cdots & r(M-2) \\ \vdots & \vdots & & \vdots \\ r^*(M-1) & r^*(M-2) & \cdots & r(0) \end{bmatrix} \tag{2.36}$$

with

$$\Delta_{M-1} = |T_{M-1}| > 0 \tag{2.37}$$

and from (2.33) using Cramer's rule [13] for the last entry in the unknown **a** vector we have

$$\delta_{M-1} = \frac{\Delta_{M-1}}{\Delta_{M-2}}. \tag{2.38}$$

Moreover (2.33) together with (2.34) gives [1]

$$H(z) = \left[z^{-(M-1)}, z^{-(M-2)}, \cdots, z^{-1}, 1 \right] \begin{bmatrix} a_{M-1} \\ \vdots \\ a_1 \\ 1 \end{bmatrix}$$

$$= \left[z^{-(M-1)}, z^{-(M-2)}, \cdots, z^{-1}, 1 \right] T_{M-1}^{-1} \begin{bmatrix} 0 \\ \vdots \\ 0 \\ \delta_{M-1} \end{bmatrix}$$

---

[1]

$$\left| \begin{matrix} A & B \\ C & D \end{matrix} \right| = |A| \, |D - CA^{-1}B|$$

- 24 -

$$
= \frac{-1}{\Delta_{M-1}}
\begin{vmatrix}
 & & & 0 \\
 & \mathbf{T}_{M-1} & & \vdots \\
 & & & 0 \\
 & & & \delta_{M-1} \\
z^{-(M-1)} & \cdots & 1 & 0
\end{vmatrix}
$$

$$
= \frac{-1}{\Delta_{M-2}}
\begin{vmatrix}
 & & & r(M-1) \\
 & \mathbf{T}_{M-2} & & r(M-2) \\
 & & & \vdots \\
 & & & r(1) \\
z^{-(M-1)} & \cdots & z^{-1} & 1
\end{vmatrix}
\tag{2.39}
$$

where we have expanded the first determinant along the last column and made use of (2.38).

The output power spectral density in (2.35) is taken as the linear prediction estimate $P_L(\omega)$. Using (2.39) this gives

$$
P_L(\omega) = S_x(\omega) = \frac{1}{|g_{M-1}(e^{-j\omega})|^2}
\tag{2.40}
$$

where

$$
g_{M-1}(z)
$$

$$
= \frac{1}{\sqrt{\Delta_{M-1}\Delta_{M-2}}}
\begin{vmatrix}
r(0) & r(1) & \cdots & r(M-2) & r(M-1) \\
r^*(1) & r(0) & \cdots & r(M-3) & r(M-2) \\
\vdots & \vdots & \ddots & \vdots & \vdots \\
r^*(M-2) & r^*(M-3) & \cdots & r(0) & r(1) \\
z^{M-1} & z^{M-2} & \cdots & z & 1
\end{vmatrix}
. \tag{2.41}
$$

The polynomial $g_{M-1}(z)$ has all its $(M-1)$ zeros in $|z| > 1$ and represents a stable filter. These zeros can lie close to the unit circle ($z = e^{j\omega}$, creating sharp peaks in the output spectrum. In general, the resolution capacity of the linear prediction based estimator is known to be superior to that of the Capon estimator [14]. To explain this, first we will relate these two estimators for a uniformly spaced array in a spatially stationary situation as above. In that case from (2.26) we have[1]

$$\frac{1}{P_C^{(M)}(\omega)} = \mathbf{a}^\dagger(\omega)\,\mathbf{T}_{M\text{-}1}^{-1}\,\mathbf{a}(\omega)$$

$$= -\frac{(1/M)}{\Delta_{M\text{-}1}}\left|\begin{array}{cccc} & & & 1 \\ & \mathbf{T}_{M\text{-}1} & & e^{-j\,\omega} \\ & & & \vdots \\ & & & e^{-j\,(M\text{-}1)\omega} \\ 1 & e^{j\,\omega} \quad \cdots \quad e^{j\,(M\text{-}1)\omega} & & 0 \end{array}\right|.$$

Using another well known fundamental result in matrix identity [2] [16], the above expression reduces to

$$\frac{1}{P_C^{(M)}(\omega)} = -\frac{(1/M)}{\Delta_{M\text{-}2}}\left|\begin{array}{cccc} & & & 1 \\ & \mathbf{T}_{M\text{-}2} & & e^{-j\,\omega} \\ & & & \vdots \\ & & & e^{-j\,(M\text{-}2)\omega} \\ 1 & e^{j\,\omega} \quad \cdots \quad e^{j\,(M\text{-}2)\omega} & & 0 \end{array}\right|$$

---

(2) Let $\mathbf{A}$ be an $n \times n$ matrix and $\mathbf{A}_{NW}$, $\mathbf{A}_{NE}$, $\mathbf{A}_{SW}$, $\mathbf{A}_{SE}$ denote the $(n\text{-}1)\times(n\text{-}1)$ minors formed from consecutive rows and consecutive columns in the northwest, northeast, southwest and southeast corners. Further let $\mathbf{A}_C$ denote the central $(n\text{-}2)\times(n\text{-}2)$ minor of $\mathbf{A}$. Then from a special case of an identity due to Jacobi [15],

$$\mathbf{A}_C\,|\mathbf{A}| = \mathbf{A}_{NW}\,\mathbf{A}_{SE} - \mathbf{A}_{NE}\,\mathbf{A}_{SW}$$

$$+ \frac{(1/M)}{\Delta_{M-1}\Delta_{M-2}} \left| det \begin{bmatrix} r^*(1) & r(0) & \cdots & r(M-3) & r(M-2) \\ r^*(2) & r^*(1) & \cdots & r(M-4) & r(M-3) \\ \vdots & \vdots & \ddots & \vdots & \vdots \\ r^*(M-1) & r^*(M-2) & \cdots & r^*(1) & r(0) \\ 1 & e^{j\omega} & \cdots & e^{j(M-2)\omega} & e^{j(M-1)\omega} \end{bmatrix} \right|^2$$

$$= \frac{1}{P_C^{(M-1)}(\omega)}$$

$$+ \frac{(1/M)}{\Delta_{M-1}\Delta_{M-2}} \left| det \begin{bmatrix} r(0) & r(1) & \cdots & r(M-2) & r(M-1) \\ r^*(1) & r(0) & \cdots & r(M-3) & r(M-2) \\ \vdots & \vdots & \ddots & \vdots & \vdots \\ r^*(M-2) & r^*(M-3) & \cdots & r(0) & r(1) \\ e^{-j(M-1)\omega} & e^{-j(M-2)\omega} & \cdots & e^{-j\omega} & 1 \end{bmatrix} \right|^2$$

$$= \frac{1}{P_C^{(M-1)}(\omega)} + \frac{1}{M} \frac{1}{P_L^{(M)}(\omega)} .$$

Completing the above recursion we have

$$\frac{1}{P_C^{(M)}(\omega)} = \frac{1}{M} \sum_{k=1}^{M} \frac{1}{P_L^{(k)}(\omega)} . \tag{2.42}$$

This "parallel resistor type" relationship between the Capon estimator and the linear prediction based estimator was first derived by Burg in connection with the maximum entropy estimator [17]. Thus the reciprocal of the Capon estimator is equal to the average of the reciprocals of the linear prediction based estimator obtained from the one point up to the M-point prediction error filter. The lower resolution of the Capon estimator is thus due to the averaging in (2.42) of the lowest to the highest resolution linear prediction estimators (see Fig. 2.7). The increased resolution is usually accompanied by a ripple

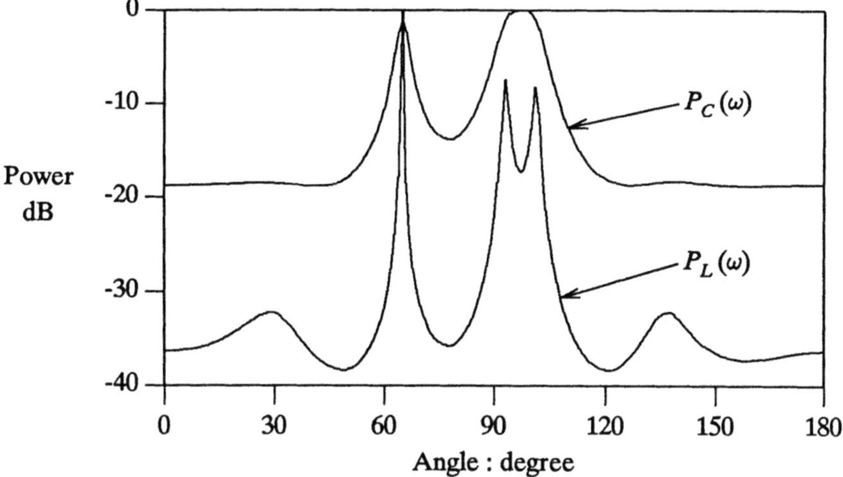

Fig. 2.7 Capon's minimum variance estimator and linear prediction based power estimator. A six-element array receives signals from three sources.

in the power estimate for the linear prediction estimator, whenever the pointing direction is away from one of the actual directions of arrival. Alternatively, the spurious peaks may be attributed to the fact that the unknown **a** vector in (2.33) and $P_L(\omega)$ make use of only the last column of $\mathbf{T}_{M-1}^{-1}$. These general statements are true when unequally spaced arrays are also involved.

The linear prediction estimator in (2.40) can be expressed in another well known form. This follows by rewritting (2.33) as

$$\mathbf{T}_{M-2} \begin{bmatrix} a_{M-1} \\ a_{M-2} \\ \vdots \\ a_1 \end{bmatrix} = - \begin{bmatrix} r(M-1) \\ r(M-2) \\ \vdots \\ r(1) \end{bmatrix} \triangleq - \gamma_{M-1,1}$$

and

$$\delta_{M-1} = r(0) + r^*(M-1)a_{M-1} + r^*(M-2)a_{M-2} + \cdots + r^*(1)a_1.$$

Let $\tilde{\mathbf{a}} = \left[a_{M-1}, a_{M-2}, \cdots, a_1 \right]^T$. Then

$$\tilde{\mathbf{a}} = -\mathbf{T}_{M-2}^{-1} \boldsymbol{\gamma}_{M-1,1} \qquad (2.43)$$

and

$$\delta_{M-1} = r(0) + \boldsymbol{\gamma}_{M-1,1}^\dagger \tilde{\mathbf{a}} = r(0) - \boldsymbol{\gamma}_{M-1,1}^\dagger \mathbf{T}_{M-2}^{-1} \boldsymbol{\gamma}_{M-1,1} . \qquad (2.44)$$

Also from (2.34)

$$H(z) = 1 + \left[z^{-(M-1)}, z^{-(M-2)}, \cdots, z^{-1} \right] \tilde{\mathbf{a}}$$

or

$$H(e^{j\omega}) = 1 + \mathbf{s}^\dagger(\omega) \tilde{\mathbf{a}} = 1 - \mathbf{s}^\dagger(\omega) \mathbf{T}_{M-2}^{-1} \boldsymbol{\gamma}_{M-1,1} \qquad (2.45)$$

where

$$\mathbf{s}(\omega) = \left[e^{j(M-1)\omega}, e^{j(M-2)\omega}, \cdots, e^{j\omega} \right]^T .$$

Thus

$$P_L(\omega) = \frac{\delta_{M-1}}{|H(e^{j\omega})|^2} = \frac{r(0) - \boldsymbol{\gamma}_{M-1,1}^\dagger \mathbf{T}_{M-2}^{-1} \boldsymbol{\gamma}_{M-1,1}}{|1 - \mathbf{s}^\dagger(\omega) \mathbf{T}_{M-2}^{-1} \boldsymbol{\gamma}_{M-1,1}|^2} . \qquad (2.46)$$

This expression will turn out to be useful in analyzing the statistical properties of the linear predictor, when data samples are used in estimating the unknown covariances.

2.3 Eigenvector-Based Techniques

Consider a linear array consisting of M identical sensors and receiving signals from K narrowband signals $u_1(t), u_2(t), \cdots, u_K(t)$ that arrive at the array from directions $\theta_1, \theta_2, \cdots, \theta_K$ with respect to the line of array (Fig. 2.8). Let $P_k; k = 1, 2, \cdots, K$ represent the signal powers and $\rho_{ij}; i,j = 1, 2, \cdots, K$ their correlation coefficients; i.e.,

$$P_k = E[\,|u_k(t)|^2\,], \quad k = 1, 2, \cdots, K \qquad (2.47)$$

and

$$\rho_{ij} = \frac{E[u_i(t) u_j^*(t)]}{\sqrt{P_i P_j}}, \quad i,j = 1, 2, \cdots, K ; \quad |\rho_{ij}| \leq 1. \qquad (2.48)$$

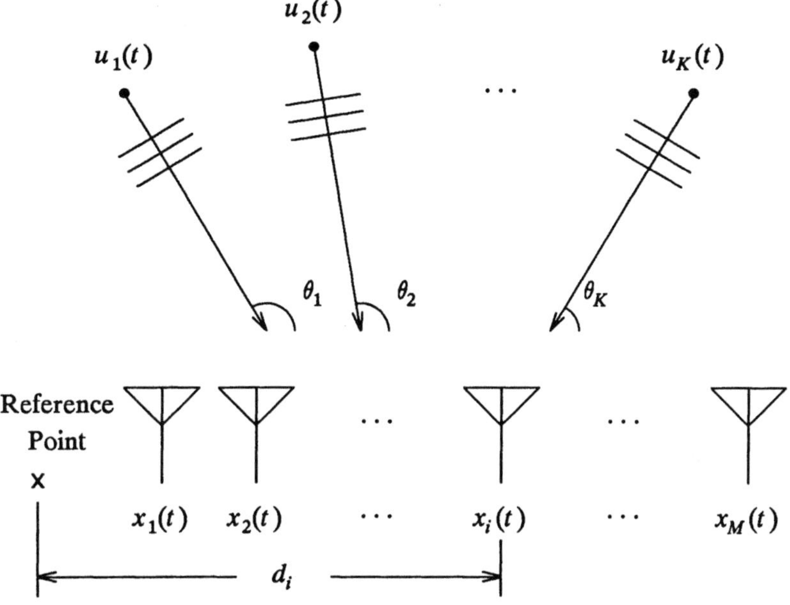

Fig. 2.8 A typical array scene.

We assume here that none of these signals are coherent (i.e., $|\rho_{ij}| \neq 1$). This case is dealt with in the next section. Using superposition of signal contributions of the form (2.13), the received signal $x_i(t)$ at the ith sensor can be written as

$$x_i(t) = \sum_{k=1}^{K} u_k(t) e^{-j\pi d_i \cos\theta_k} + n_i(t). \tag{2.49}$$

As before d_i here represents the ith sensor element position normalized to half the common wavelength, with respect to some common reference point, and $n_i(t)$ stands for the additive noise at the ith sensor. It is assumed that signals and noises are stationary, zero mean, uncorrelated random processes, and further as in (2.5) the noises are assumed to be uncorrelated and identical between themselves with common variance σ^2. Rewriting (2.49) in common vector notation and with $\omega_k = \pi\cos\theta_k$; $k = 1, 2, \cdots, K$, we have

$$\mathbf{x}(t) = \sqrt{M} \sum_{k=1}^{K} u_k(t)\,\mathbf{a}(\omega_k) + \mathbf{n}(t) \tag{2.50}$$

where $\mathbf{x}(t)$ is the $M \times 1$ vector

$$\mathbf{x}(t) = \left[x_1(t), x_2(t), \cdots, x_M(t) \right]^T \tag{2.51}$$

in an M dimensional Hilbert space H over the complex field C, and $\mathbf{a}(\omega_k)$ is the normalized direction vector associated with the arrival angle θ_k; i.e.,

$$\mathbf{a}(\omega_k) = \frac{1}{\sqrt{M}} \left[e^{-j d_1 \omega_k}, e^{-j d_2 \omega_k}, \cdots, e^{-j d_M \omega_k} \right]^T . \tag{2.52}$$

The array output vector $\mathbf{x}(t)$ can further be written as

$$\mathbf{x}(t) = \mathbf{A}\,\mathbf{u}(t) + \mathbf{n}(t), \tag{2.53}$$

where

$$\mathbf{u}(t) = \left[u_1(t), u_2(t), \cdots, u_K(t) \right]^T \tag{2.54}$$

$$\mathbf{n}(t) = \left[n_1(t), n_2(t), \cdots, n_M(t) \right]^T$$

and

$$\mathbf{A} = \sqrt{M} \left[\mathbf{a}(\omega_1), \mathbf{a}(\omega_2), \cdots, \mathbf{a}(\omega_K) \right] \tag{2.55}$$

is an $M \times K$ matrix consisting of K direction vectors. From our assumptions it follows that the $M \times M$ array output covariance matrix

$$\mathbf{R} = E\left[\mathbf{x}(t)\mathbf{x}^\dagger(t) \right] \tag{2.56}$$

has the form

$$\mathbf{R} = \mathbf{A} E\left[\mathbf{u}(t)\mathbf{u}^\dagger(t) \right] \mathbf{A}^\dagger + E\left[\mathbf{n}(t)\mathbf{n}^\dagger(t) \right] = \mathbf{A} \mathbf{R}_u \mathbf{A}^\dagger + \sigma^2 \mathbf{I}. \tag{2.57}$$

Notice that $\mathbf{R}$ is hermitian and positive definite and by definition

$$\mathbf{R}_u = E\left[\mathbf{u}(t)\mathbf{u}^\dagger(t) \right] \tag{2.58}$$

represents the source covariance matrix of size $K \times K$ which remains nonsingular so long as there are no coherent sources present in the data ($|\rho_{ij}| \neq 1$). In that case $\mathbf{R}_u$ is of rank K and from the linear independence of the direction vectors $\mathbf{a}(\omega_1)$, $\mathbf{a}(\omega_2)$, $\cdots$, $\mathbf{a}(\omega_K)$, $\mathbf{A}$ is also of rank K, which taken together implies that the nonnegative definite matrix $\mathbf{A}\mathbf{R}_u\mathbf{A}^\dagger$ is also of rank K. Let $\mu_1 \geq \mu_2 \geq \cdots \geq \mu_K$ denote its K nonzero eigenvalues. Then the M eigenvalues of R are given by

$$\lambda_k = \begin{cases} \mu_k + \sigma^2 & k = 1, 2, \cdots, K \\ \sigma^2 & k = K+1, K+2, \cdots, M \end{cases} \tag{2.59}$$

and further let β_1, β_2, $\cdots$, β_K, $\cdots$, β_M, represent their associated eigenvectors. Then

$$\mathbf{R} = \sum_{k=1}^{M} \lambda_k\, \beta_k\, \beta_k^\dagger = \mathbf{B}\mathbf{A}\mathbf{B}^\dagger \tag{2.60}$$

where

$$\mathbf{B} = \left[\beta_1, \beta_2, \cdots, \beta_K, \beta_{K+1}, \cdots, \beta_M \right], \tag{2.61}$$

$$\mathbf{B}\mathbf{B}^\dagger = \mathbf{I}_M \tag{2.62}$$

and

$$\mathbf{A} = diag[\, \lambda_1, \lambda_2, \cdots, \lambda_K, \sigma^2, \sigma^2, \cdots, \sigma^2\,]. \tag{2.63}$$

Moreover, for any $i > K$, from (2.59)

$$\mathbf{R}\beta_i = \lambda_i\,\beta_i = \sigma^2\beta_i\,.$$

But

$$\mathbf{R}\beta_i = (\mathbf{A}\mathbf{R}_u\mathbf{A}^\dagger + \sigma^2\mathbf{I})\beta_i$$

and together they imply

$$\mathbf{A}\mathbf{R}_u\mathbf{A}^\dagger\beta_i = 0$$

or

$$\mathbf{A}^\dagger\beta_i = 0 \quad \Longleftrightarrow \quad \beta_i^\dagger\mathbf{a}(\omega_k) = 0, \quad K+1 \leq i \leq M, 1 \leq k \leq K\,. \tag{2.64}$$

The last step follows from the full rank property of $\mathbf{A}$ and $\mathbf{R}$. Stated in words, the eigenvectors associated with the repeating lowest eigenvalue of $\mathbf{R}$ are orthogonal to the direction vectors corresponding to the actual angles of arrival. This remarkable observation forms the cornerstone for almost all eigenvector-based algorithms [18 - 27, 30, 33, 43 - 48, 50]. To view this geometrically, notice that the (orthonormal) eigenvectors β_{K+1}, β_{K+2}, $\cdots$, β_M, associated with the lowest eigenvalue (noise variance) σ^2, together span an $(M\text{-}K)$ dimensional subspace N of H and from (2.64) every direction vector associated with an actual arrival angle is orthogonal to this subspace. Since the K direction vectors $\mathbf{a}(\omega_1)$, $\mathbf{a}(\omega_2)$, $\cdots$, $\mathbf{a}(\omega_K)$ are themselves linearly independent, the K dimensional "signal" subspace S spanned by these actual direction vectors is orthogonal to the subspace N. Thus N is the orthogonal complement of S and may be fittingly designated as the "noise" subspace. Moreover, since H can always be written as the direct sum of any finite dimensional subspace and its orthogonal complement [28], we have

$$H = S \oplus N \tag{2.65}$$

where

$$S = span\{\, \mathbf{a}(\omega_1), \mathbf{a}(\omega_2), \cdots, \mathbf{a}(\omega_K)\,\}$$

and

$$N = span\{\, \beta_{K+1}, \beta_{K+2}, \cdots, \beta_M\,\}\,.$$

Thus the peaks of the function

$$P(\omega) = \frac{1}{\displaystyle\sum_{i=K+1}^{M} |\beta_i^{\dagger}\mathbf{a}(\omega)|^2} \tag{2.66}$$

or equivalently the K zeros of the function

$$Q(\omega) = \sum_{i=K+1}^{M} |\beta_i^{\dagger}\mathbf{a}(\omega)|^2 \tag{2.67}$$

will correspond to the true directions of arrival. Following Schmidt [19], we will refer to this procedure as the MUltiple SIgnal Classification (MUSIC) technique. Notice that for (2.66) to be meaningful $M \geq K+1$ and this restricts the minimum number of required sensor elements to at least one more than the total number of sources present in the scene. Unlike the estimators discussed in the previous section, $P(\omega)$ here does not estimate the signal power associated with

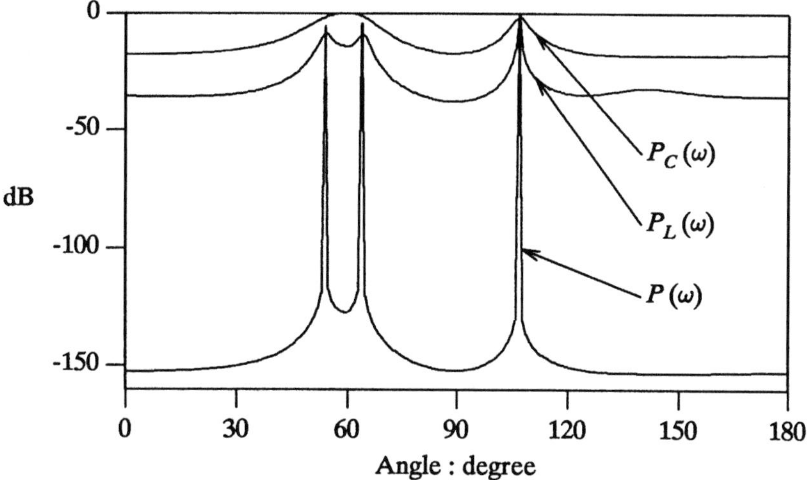

Fig. 2.9 MUSIC estimator $P(\omega)$ *vs.* Capon's minimum variance estimator and linear prediction estimator. A five-element array receives signals from three sources.

each arrival angle. Instead, when the ensemble average **R** of the array output covariance matrix is known exactly, under uncorrelated and identical noise conditions, the peaks of $P(\omega)$ are guaranteed to correspond to the true angles of arrival. Since these peaks (or zeros of $Q(\omega)$) are always distinct irrespective of the actual separation between arrival angles, in principle these estimators can distinguish and resolve arbitrarily closely situated targets (see Fig. 2.9).

However, when some of the signals present in (2.49) are perfectly correlated (i.e., coherent), as happens, for example in multipath propagation, $\mathbf{R}_u$ in (2.58) becomes singular, and the above conclusions are no longer true. Several alternatives have been proposed [23 - 27] to deal with this situation and before examining them, we will establish an interesting nonambiguity property for the array output covariance matrix in a completely coherent situation.

2.3.1 Completely Coherent Case

Let K narrow band coherent signals arrive at the array (described as above) from directions $\theta_1, \theta_2, \cdots, \theta_K$. At any instant these K signals $u_1(t), u_2(t), \cdots, u_K(t)$ are phase delayed, amplitude

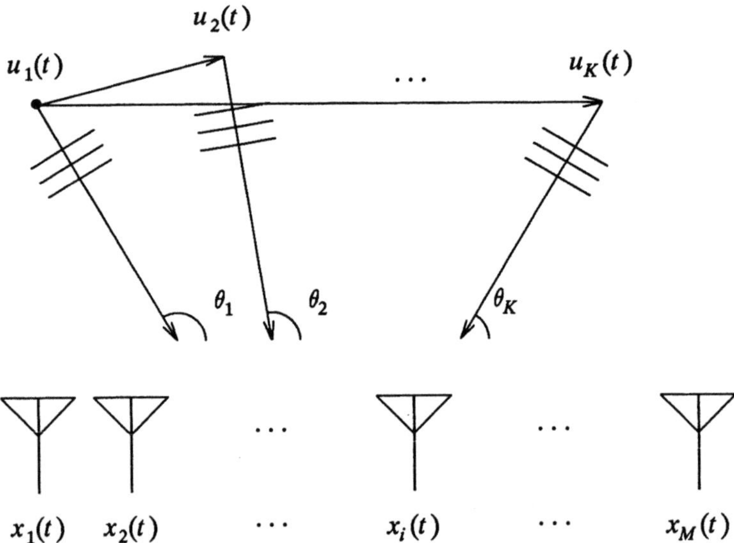

Fig. 2.10 A completely coherent situation.

weighted replicas of one of them − say the first − and hence (see Fig. 2.10)

$$u_k(t) = \alpha_k\, u_1(t)\,;\quad k = 1, 2, \cdots, K\,;\quad \alpha_1 = 1 \qquad (2.68)$$

where α_k represents the complex attenuation of the kth signal with respect to the first signal, $u_1(t)$. With $\alpha_k = \rho_k\, e^{j\phi_k}; k = 1, 2, \cdots, K$, ρ_k is the amplitude attenuation and ϕ_k the relative phase delay of the kth signal with reference to the first one. From (2.50), the array output vector $\mathbf{x}(t)$ can be written as

$$\mathbf{x}(t) = \sqrt{M}\, u_1(t) \sum_{k=1}^{K} \alpha_k\, \mathbf{a}(\omega_k) + \mathbf{n}(t) = u_1(t)\mathbf{b} + \mathbf{n}(t) \qquad (2.69)$$

where

$$\mathbf{b} = \sqrt{M} \sum_{k=1}^{K} \alpha_k\, \mathbf{a}(\omega_k). \qquad (2.70)$$

Further the uncorrelated sensor noises are allowed to have unequal variance; i.e.,

$$E[n_i(t)n_j^*(t)] = \sigma_i^2 \delta_{ij} . \tag{2.71}$$

With (2.71) and (2.69) in (2.56), and under the normalization that $E[\,|u_1(t)|^2\,] = 1$, the array output covariance matrix $\mathbf{R}$ becomes

$$\mathbf{R} = \mathbf{b}\,\mathbf{b}^\dagger + \begin{bmatrix} \sigma_1^2 & & & \mathbf{O} \\ & \sigma_2^2 & & \\ & & \ddots & \\ \mathbf{O} & & & \sigma_M^2 \end{bmatrix} . \tag{2.72}$$

Youla has shown that [29], for a given $\mathbf{R}$ of the form (2.72), the $\mathbf{b}$ vector (up to multiplication by a phase factor) and the noise variances σ_i^2; $i = 1, 2, \cdots, M$ are always unique, provided the array has at least three sensors ($M \geq 3$).

To prove this, let us assume the contrary, whence $\mathbf{R}$ has two such representations; i.e.,

$$\mathbf{R} = \mathbf{b}\,\mathbf{b}^\dagger + \begin{bmatrix} \sigma_1^2 & & & \mathbf{O} \\ & \sigma_2^2 & & \\ & & \ddots & \\ \mathbf{O} & & & \sigma_M^2 \end{bmatrix} = \mathbf{c}\,\mathbf{c}^\dagger + \begin{bmatrix} \delta_1^2 & & & \mathbf{O} \\ & \delta_2^2 & & \\ & & \ddots & \\ \mathbf{O} & & & \delta_M^2 \end{bmatrix} \tag{2.73}$$

and this gives

$$\mathbf{b}\,\mathbf{b}^\dagger - \mathbf{c}\,\mathbf{c}^\dagger = \begin{bmatrix} \delta_1^2 - \sigma_1^2 & & & \mathbf{O} \\ & \delta_2^2 - \sigma_2^2 & & \\ & & \ddots & \\ \mathbf{O} & & & \delta_M^2 - \sigma_M^2 \end{bmatrix} . \tag{2.74}$$

Notice that in the above expression, the left-hand side is the difference of two rank one matrices and is at most of rank two. Thus the right-hand side is also at most of rank two, and as a result $(M\text{-}2)$ of its entries must be zeros. Without loss of generality assume that $\delta_3^2 = \sigma_3^2$, $\delta_4^2 = \sigma_4^2, \cdots, \delta_M^2 = \sigma_M^2$ (if not, rearrange the elements). Thus (2.74) reduces to

$$
\mathbf{b}\mathbf{b}^\dagger - \mathbf{c}\mathbf{c}^\dagger =
\begin{bmatrix}
\delta_1^2 - \sigma_1^2 & & & & \mathbf{O} \\
& \delta_2^2 - \sigma_2^2 & & & \\
& & 0 & & \\
& & & \ddots & \\
\mathbf{O} & & & & 0
\end{bmatrix} .
\tag{2.75}
$$

To proceed further, partition $\mathbf{b}$ and $\mathbf{c}$ as follows

$$
\mathbf{b} =
\begin{bmatrix}
\mathbf{b}_1 \\
\text{---} \\
\mathbf{b}_2
\end{bmatrix}
\begin{matrix} 2 \\ \\ M\text{-}2 \end{matrix}
\quad ; \quad
\mathbf{c} =
\begin{bmatrix}
\mathbf{c}_1 \\
\text{---} \\
\mathbf{c}_2
\end{bmatrix}
\begin{matrix} 2 \\ \\ M\text{-}2 \end{matrix}
\tag{2.76}
$$

where $\mathbf{b}_1$, $\mathbf{c}_1$ are of size 2×1 and $\mathbf{b}_2$, $\mathbf{c}_2$ are of size $(M\text{-}2)\times 1$. With (2.76) in (2.75) we have

$$
\begin{bmatrix} \mathbf{b}_1 \\ \mathbf{b}_2 \end{bmatrix}
\begin{bmatrix} \mathbf{b}_1^\dagger & \mathbf{b}_2^\dagger \end{bmatrix}
-
\begin{bmatrix} \mathbf{c}_1 \\ \mathbf{c}_2 \end{bmatrix}
\begin{bmatrix} \mathbf{c}_1^\dagger & \mathbf{c}_2^\dagger \end{bmatrix}
$$

$$
=
\begin{bmatrix}
\delta_1^2 - \sigma_1^2 & & & & \mathbf{O} \\
& \delta_2^2 - \sigma_2^2 & & & \\
& & 0 & & \\
& & & \ddots & \\
\mathbf{O} & & & & 0
\end{bmatrix}
\tag{2.77}
$$

which gives

$$b_1 b_2^\dagger - c_1 c_2^\dagger = O, \qquad (2.78)$$

$$b_2 b_2^\dagger - c_2 c_2^\dagger = O. \qquad (2.79)$$

Since by assumption the array has at least three sensors, $\mathbf{b}$ has at least three nonzero entries and hence $\mathbf{b}_2$ and $\mathbf{c}_2$ cannot be zero vectors; i.e., $\mathbf{b}_2 \neq 0$, $\mathbf{c}_2 \neq 0$. Multiplying (2.78) and (2.79) by $\mathbf{c}_2$ from the right, we obtain

$$b_1(b_2^\dagger c_2) - c_1(c_2^\dagger c_2) = 0.$$
$$b_2(b_2^\dagger c_2) - c_2(c_2^\dagger c_2) = 0$$

or

$$b_1 = \mu c_1 \qquad (2.80)$$

$$b_2 = \mu c_2 \qquad (2.81)$$

where

$$\mu = c_2^\dagger c_2 / b_2^\dagger c_2.$$

With (2.81) in (2.79) we also have $(1 - |\mu|^2)c_2^\dagger c_2 = 0$, which together with $c_2 \neq 0$ gives $1 - |\mu|^2 = 0$ or $|\mu| = 1$. Finally from (2.77) and (2.80) we have

$$b_1 b_1^\dagger - c_1 c_1^\dagger = (1 - |\mu|^2)c_1 c_1^\dagger = O = \begin{bmatrix} \delta_1^2 - \sigma_1^2 & 0 \\ 0 & \delta_2^2 - \sigma_2^2 \end{bmatrix}$$

and hence $\delta_1^2 = \sigma_1^2$ and $\delta_2^2 = \sigma_2^2$. Thus $\mathbf{b} = \mu c$, $|\mu| = 1$ and $\delta_i^2 = \sigma_i^2$; $i = 1, 2, \cdots, M$, proving our claim. To summarize, the decomposition $\mathbf{R} = \mathbf{b}\mathbf{b}^\dagger + diag[\sigma_1^2, \sigma_2^2, \cdots, \sigma_M^2]$ is unique, if the array has at least three nonzero outputs, and this is almost always satisfied whenever $M \geq 3$.

This raises the interesting question: If $\mathbf{b}$ and σ_i^2; $i = 1, 2, \cdots, M$ are unique, how can they and in particular the arrival angles θ_1, θ_2, $\cdots$, θ_K be evaluated?

To answer this, notice that $\mathbf{R} - diag[\sigma_1^2, \sigma_2^2, \cdots, \sigma_M^2]$ is a rank one matrix and hence all 2×2 and higher order minors of

$$
\begin{bmatrix}
r_{11}-\sigma_1^2 & r_{12} & r_{13} & \cdots & r_{1M} \\
r_{12}^* & r_{22}-\sigma_2^2 & r_{23} & \cdots & r_{2M} \\
r_{13}^* & r_{23}^* & r_{33}-\sigma_3^2 & \cdots & r_{3M} \\
\vdots & \vdots & \vdots & \ddots & \vdots \\
r_{1M}^* & r_{2M}^* & r_{3M}^* & \cdots & r_{MM}-\sigma_M^2
\end{bmatrix}
$$

are zeros. If M is an odd number $(M = 2m+1)$, then the order of the largest minor containing only one unknown σ_i^2; $i = 1, 2, \cdots, M$ is m and using these the noise variances can be determined. For example with $M = 3$ we have

$$
\sigma_1^2 = \frac{r_{11}r_{23} - r_{12}^* r_{13}}{r_{23}}, \quad \sigma_2^2 = \frac{r_{22}r_{13} - r_{12}r_{23}}{r_{13}},
$$

and

$$
\sigma_3^2 = \frac{r_{33}r_{12} - r_{23}^* r_{13}}{r_{12}}.
$$

Then $\mathbf{R}_1 = \mathbf{R} - diag[\sigma_1^2, \sigma_2^2, \cdots, \sigma_M^2]$ is of unit rank and has the form $\mathbf{b}\mathbf{b}^\dagger$. Since $\mathbf{R}_1\mathbf{b} = \mathbf{b}(\mathbf{b}^\dagger\mathbf{b}) = \|\mathbf{b}\|^2\mathbf{b}$, the eigenvector $\boldsymbol{\beta}_1$ corresponding to the nonzero eigenvalue of $\mathbf{R}_1$ is proportional to $\mathbf{b}$. It is easily verified that $\mathbf{b} = \|\mathbf{b}\|\boldsymbol{\beta}_1$ with $\|\mathbf{b}\|^2 = tr(\mathbf{R}_1)$. Thus $\mathbf{b}$ is known, and the actual angles of arrival in (2.69) can be obtained from this vector by employing a special case of the technique discussed in the next section. It may be pointed out that in the above discussion all sources are assumed to be completely coherent with each other. Next, we deal with the most general source situation: a correlated source environment consisting of coherent as well as partially correlated signals in presence of noise.

2.3.2 Symmetric Array Scheme: Coherent Sources in a Correlated Scene

In this case the source scene consists of $K+J$ narrowband signals $u_1(t)$, $u_2(t)$, $\cdots$, $u_K(t)$, $u_{K+1}(t)$, $\cdots$, $u_{K+J}(t)$, of which the first K signals are completely coherent and the last $J+1$ signals are partially correlated. Thus the coherent signals are partially correlated with the remaining set of signals. Further, the respective arrival angles are assumed to be θ_1, θ_2, $\cdots$, θ_K, θ_{K+1}, $\cdots$, θ_{K+J}. These $(K+J)$ signals are received at an M element linear array with $M > K+J$, where one of the end elements is taken as the reference point. Using (2.68), this allows us to write the reference element output $x_0(t)$ at time t as

$$x_0(t) = \sum_{k=1}^{K+J} u_k(t) + n_0(t) = u_1(t) \sum_{k=1}^{K} \alpha_k + \sum_{k=K+1}^{K+J} u_k(t) + n_0(t)$$

$$= u_1(t)b_0 + \sum_{k=K+1}^{K+J} u_k(t) + n_0(t) \tag{2.82}$$

where $b_0 = \sum_{k=1}^{K} \alpha_k \neq 0$. Further, the ith element output can be written as

$$x_i(t) = u_1(t) \sum_{k=1}^{K} \alpha_k e^{-j\pi d_i \cos\theta_k} + \sum_{k=K+1}^{K+J} u_k(t) e^{-j\pi d_i \cos\theta_k} + n_i(t),$$

$$i = 1, 2, \cdots, M\text{-}1. \tag{2.83}$$

Barring the reference element, let $n_i(t)$ represent spatially correlated noise of unequal variances

$$E[n_i(t)n_j^*(t)] = \Sigma_{ij} \neq 0, \quad i,j = 1, 2, \cdots, M\text{-}1$$

and $E[n_0(t)n_i^*(t)] = \sigma_0^2 \delta_{0i}$. Our main objective here is to determine the unknown directions of arrival θ_1, θ_2, $\cdots$, θ_K, $\cdots$, θ_{K+J} from the array output covariances. In what follows, we demonstrate a scheme that is applicable for noise fields of unequal intensity at different array elements. Toward this purpose define

$$r(i) = E[x_0(t)x_i^*(t)], \quad i = 0, 1, \cdots, M\text{-}1.$$

Then using (2.82) - (2.83) and (2.47) - (2.48) we have

$$r(i) = \sum_{k=1}^{K} \left[P_1 b_0 \alpha_k^* + \sum_{l=K+1}^{K+J} \sqrt{P_1 P_l}\, \rho_{1l}^* \alpha_k^* \right] e^{j\pi d_i \cos\theta_k}$$

$$+ \sum_{k=K+1}^{K+J} \left[\sqrt{P_1 P_k}\, \rho_{k1}^* b_0 + \sum_{l=K+1}^{K+J} \sqrt{P_k P_l}\, \rho_{kl}^* \right] e^{j\pi d_i \cos\theta_k} + \sigma_0^2 \delta_{0i}$$

$$= \sum_{k=1}^{K+J} b_k\, e^{j\pi d_i \cos\theta_k} + \sigma_0^2 \delta_{0i}, \quad i = 0, 1, \cdots, M\text{-}1 \qquad (2.84)$$

where

$$b_k = \begin{cases} \sqrt{P_1}\left[\sqrt{P_1}\, b_0 + \displaystyle\sum_{l=K+1}^{K+J} \sqrt{P_l}\, \rho_{1l}^* \right] \alpha_k^* & k = 1, 2, \cdots, K \\[2em] \sqrt{P_k}\left[\sqrt{P_1}\, \rho_{k1}^* b_0 + \displaystyle\sum_{l=K+1}^{K+J} \sqrt{P_l}\, \rho_{kl}^* \right] & k = K+1, \cdots, K+J. \end{cases} \qquad (2.85)$$

Notice that the constants, b_k, in general, are complex numbers (except in special cases). Had all b_k; $k = 1, 2, \cdots, K+J$ been real, then in the case of a *uniform array* (i.e., $d_i = i$), the hermitian Toeplitz matrix

$$\mathbf{T}_0 = \begin{bmatrix} r(0) & r(1) & \cdots & r(M\text{-}1) \\ r^*(1) & r(0) & \cdots & r(M\text{-}2) \\ \vdots & \vdots & \ddots & \vdots \\ r^*(M\text{-}1) & r^*(M\text{-}2) & \cdots & r(0) \end{bmatrix} \qquad (2.86)$$

generated by $r(0), r(1), \cdots, r(M\text{-}1)$ has some interesting structural properties [30]. In fact, it follows readily from (2.84) that

$$\mathbf{T}_0 = \tilde{\mathbf{A}}\,\mathbf{B}\,\tilde{\mathbf{A}}^\dagger + \sigma_0^2 \mathbf{I}_M \qquad (2.87)$$

where $\tilde{\mathbf{A}}$ is an $M \times (K+J)$ Vandermonde matrix given by

$$\tilde{\mathbf{A}} = \sqrt{M} \left[\mathbf{a}(\omega_1), \mathbf{a}(\omega_2), \cdots, \mathbf{a}(\omega_K), \cdots, \mathbf{a}(\omega_{K+J}) \right], \qquad (2.88)$$

$$\mathbf{B} = \begin{bmatrix} b_1 & & & & & \\ & b_2 & & & \mathbf{O} & \\ & & \ddots & & & \\ & & & b_K & & \\ & \mathbf{O} & & & \ddots & \\ & & & & & b_{K+J} \end{bmatrix} \qquad (2.89)$$

and $\mathbf{a}(\omega_k)$; $k = 1, 2, \cdots, K+J$ are the normalized direction vectors associated with the distinct arrival angles for a uniform array with its first element serving as the reference element.

Since $(M \geq K+J)$ the direction vectors are independent, $\tilde{\mathbf{A}}$, and therefore $\tilde{\mathbf{A}} \mathbf{B} \tilde{\mathbf{A}}^\dagger$ are both of rank $(K+J)$. This in turn implies that $(M-K-J)$ eigenvalues of $\mathbf{T}_0$ are σ_0^2 and further as in section 2.3, the corresponding eigenvectors $\boldsymbol{\beta}_{(1)}, \boldsymbol{\beta}_{(2)}, \cdots, \boldsymbol{\beta}_{(M-K-J)}$ are orthogonal to the actual direction vectors $\mathbf{a}(\omega_1), \mathbf{a}(\omega_2), \cdots, \mathbf{a}(\omega_{K+J})$: i.e.,

$$\boldsymbol{\beta}_{(i)}^\dagger \mathbf{a}(\omega_k) = 0, \quad k = 1, 2, \cdots, K+J, \, i = 1, 2, \cdots, M-K-J.$$

Once again the $K+J$ zeros of the function

$$Q_0(\omega) = \sum_{i=1}^{M-K-J} |\boldsymbol{\beta}_{(i)}^\dagger \mathbf{a}(\omega)|^2 \qquad (2.90)$$

will correspond to the true arrival angles. This, however, raises the following question: Can some suitable preprocessing be done on the array outputs to transform b_k; $k = 1, 2, \cdots, K+J$ in (2.89) to real? This can be answered affirmatively by extending the M element array to incorporate a new set of M-1 elements symmetrically about the reference point (see Fig. 2.11). Then the output at the ith element of this new set of elements will be

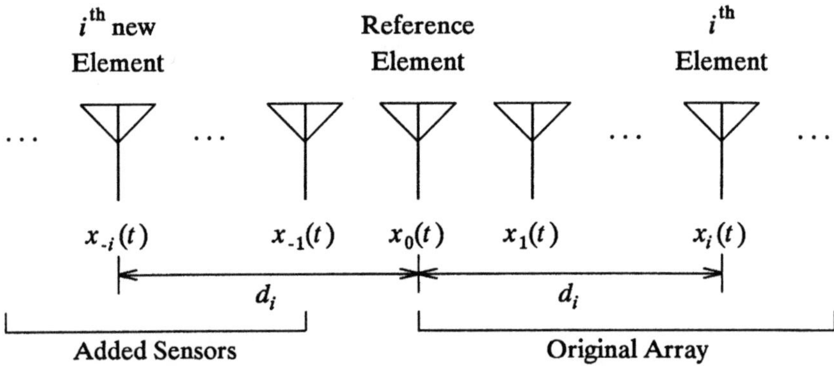

Fig. 2.11 Sensor array placement for decoherency.

$$x_{-i}(t) = u_1(t) \sum_{k=1}^{K} \alpha_k e^{j\pi d_1 \cos\theta_k} + \sum_{k=K+1}^{K+J} u_k(t) e^{j\pi d_1 \cos\theta_k} + n_{-i}(t) \quad (2.91)$$

As before define

$$\bar{r}(i) \triangleq E[x_0(t)x_{-i}^*(t)], \quad i = 1, 2, \cdots, M\text{-}1.$$

Then it is easy to show that

$$\bar{r}(i) = \sum_{k=1}^{K+J} b_k e^{-j\pi d_1 \cos\theta_k}, \quad i = 1, 2, \cdots, M\text{-}1. \quad (2.92)$$

where $b_k, k = 1, 2, \cdots, K+J$ are as in (2.85). Now let

$$R(i) = [r(i) + \bar{r}^*(i)]/2, \quad i = 1, 2, \cdots, M\text{-}1. \quad (2.93)$$

Using (2.84) and (2.92) in (2.93) we have

$$R(i) = \sum_{k=1}^{K+1} c_k e^{j\pi d_1 \cos\theta_k}, \quad i = 1, 2, \cdots, M\text{-}1 \quad (2.94)$$

where by definition

$$c_k = Re[b_k], \quad k = 1, 2, \cdots, K+J. \quad (2.95)$$

This together with $r(0)$ given by (2.84) allows us to define

$$R(0) = Re[r(0)] = \sum_{k=1}^{K+J} Re[b_k] + \sigma_0^2 = \sum_{k=1}^{K+J} c_k + \sigma_0^2. \quad (2.96)$$

Then the hermitian Toeplitz matrix $\mathbf{T}_s$ formed from these $R(i)$; $i = 0$, 1, 2, $\cdots$, M-1 has the form

$$\mathbf{T}_s = \begin{bmatrix} R(0) & R(1) & \cdots & R(M\text{-}1) \\ R^*(1) & R(0) & \cdots & R(M\text{-}2) \\ \vdots & \vdots & \ddots & \vdots \\ R^*(M\text{-}1) & R^*(M\text{-}2) & \cdots & R(0) \end{bmatrix} \quad (2.97)$$

and can be written as

$$\mathbf{T}_s = \tilde{\mathbf{A}}\mathbf{C}\tilde{\mathbf{A}}^\dagger + \sigma_0^2 \mathbf{I}_M \quad (2.98)$$

where $\tilde{\mathbf{A}}$ is as defined in (2.88) and

$$\mathbf{C} = \begin{bmatrix} c_1 & & & & & \\ & c_2 & & & \mathbf{O} & \\ & & \ddots & & & \\ & & & c_K & & \\ & \mathbf{O} & & & \ddots & \\ & & & & & c_{K+J} \end{bmatrix}. \quad (2.99)$$

Note that c_k; $k = 1, 2, \cdots, K+J$ being real, $\tilde{\mathbf{A}}\mathbf{C}\tilde{\mathbf{A}}^\dagger$ is of rank $K+J$, and from the discussion that follows (2.89), it is clear that the procedure in (2.90) can be applied to estimate the actual angles of arrival. In that case the $K+J$ zeros of $Q_s(\omega)$ given by

$$Q_s(\omega) = \sum_{i=1}^{M-K-J} |\beta_{(i)}^\dagger \mathbf{a}(\omega)|^2 \quad (2.100)$$

will correspond to the true directions of arrival $\theta_1, \theta_2, \cdots, \theta_k, \cdots, \theta_{K+J}$.

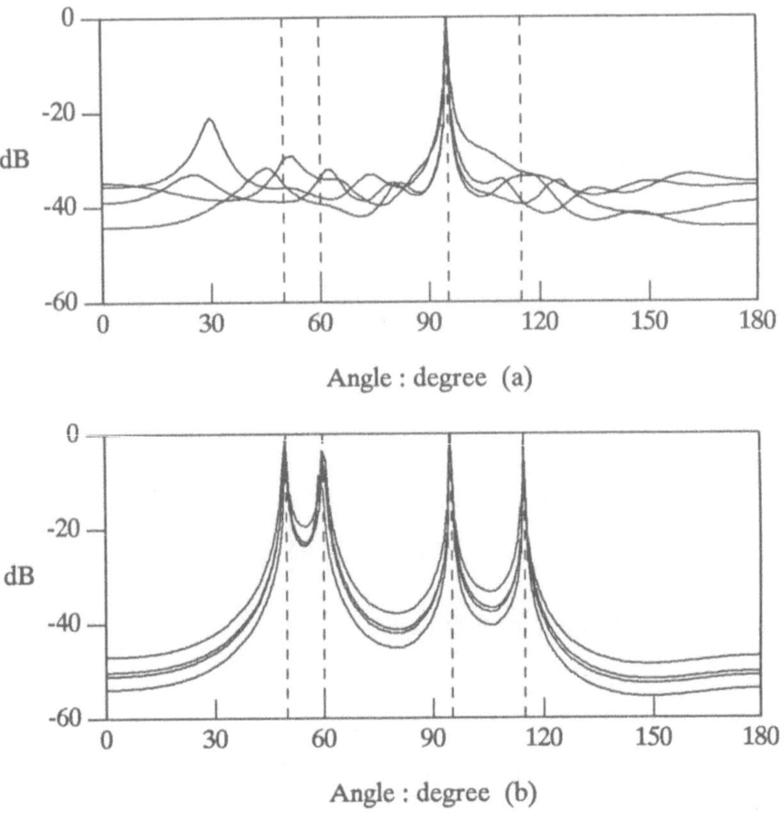

Fig. 2.12 Spatial spectrum for a mixed scenario. (a) $P(\omega)$ using the conventional MUSIC algorithm. (b) $P(\omega) = 1/Q_s(\omega)$ using the symmetric array scheme.

Fig. 2.12 shows results of simulation in a mixed source scene using the above procedure. A nine-element uniform array receives signals from three coherent sources and an uncorrelated source located along 50°, 60°, 115°, and 95° respectively in presence of noise of equal variances at different elements. The multipath coefficients of the coherent sources are taken to be $\alpha_1 = 1.0$, $\alpha_2 = -0.45+j0.34$, and $\alpha_3 = 0.29+j0.2$. The input SNRs of the reference signal for the coherent set and the uncorrelated signal are 10 and 7 dBs respectively. Six hundred data samples are used to compute the maximum likelihood estimates of the covariances using (3.7). Results of computation using the conventional MUSIC algorithm on the 9×9 array output covariance matrix as well as the symmetric array scheme described above are included here. In the latter case a 5×5 matrix representing $\mathbf{T}_s$ in (2.97) is constructed and is used in accordance with (2.100). Clearly, the directions of all three multipath arrivals are resolvable in this case. The improvement in terms of resolvability and dynamic range of signal peaks to background noise levels is also visible for the new technique.

Although the symmetric array based scheme can successfully deal with arbitrary source correlations and arbitrary sensor noise fields (except the reference element), for its implementation, a large number of sensor elements are required compared to the conventional case. In particular, notice that for (2.90) to be meaningful $M\text{-}K\text{-}J \geq 1$, requiring the total number of sensor elements $M_0 = 2M\text{-}1$ to satisfy $M_0 > 2(K+J)$. Thus in a completely coherent situation ($J = 0$), the totality of sensors must be at least one more than twice the number of sources present in the scene. However, this additional requirement is not always necessary and can be eliminated in the special case of symmetric multipath, where every coherent arrival may have arbitrary path attenuation, but all of them have the same phase delay. In this case, as shown below, the array need not have symmetry and further only $K+1$ uniformly placed sensors are required to estimate any such K coherent arrivals.

Symmetric Multipath Case

Symmetric multipath refers to the case where the phase delays ϕ_k in $\alpha_k = \rho_k\, e^{j\,\phi_k}$ in (2.68) are all of the same value. This situation is particularly relevant when there are two sources ($K = 2$, one direct and another indirect), with equal arrival angles [31]. In what follows, we consider the general situation of K symmetric multipath (with $K \geq 2$) with arbitrary directions of arrival.

In this case, with $\phi_k = \phi; k = 1, 2, \cdots, K$, with respect to a suitable reference point, α_k in (2.68) reduces to

$$\alpha_k = \rho_k\, e^{j\,\phi}, \quad k = 1, 2, \cdots, K \qquad (2.101)$$

and hence

$$b_0 = \sum_{k=1}^{K} \alpha_k = \left(\sum_{k=1}^{K} \rho_k \right) e^{j\,\phi} = \rho\, e^{j\,\phi}$$

where

$$\rho = \sum_{k=1}^{K} \rho_k > 0.$$

Thus from (2.84) and (2.85) with $P_1 = 1$ and $J = 0$, we have

$$r(0) = |b_0|^2 + \sigma_0^2 = \rho^2 + \sigma_0^2 > 0$$

and from (2.85)

$$b_k = b_0 \alpha_k^* = \rho \rho_k > 0.$$

Thus $b_k; k = 1, 2, \cdots, K$ are all *positive real* (p.r.) numbers. As a result, the hermitian Toeplitz matrix $\mathbf{T}_0$ in (2.86), formed by $r(0)$, $r(1)$, $\cdots$, $r(M\text{-}1)$, is positive definite and $\tilde{\mathbf{A}}\mathbf{B}\tilde{\mathbf{A}}^\dagger$ is nonnegative definite and of rank K. Hence the lowest eigenvalue σ_0^2 will repeat $(M\text{-}K+1)$ times and the corresponding eigenvectors will be orthogonal to the true multipath directions of arrival. Note that in this case we *do not* need a symmetric array. Moreover, the noise field can be arbitrary in this case. To summarize, using a uniformly placed M element linear array, one can resolve up to $(M\text{-}1)$ symmetric coherent directions of arrival by making use of the procedure discussed above.

The symmetric array scheme described above does not make use of all available output covariances in its implementation. In fact, it only uses the first row (or column) of the array output covariance matrix. This is underutilization of the available information especially when actual data samples are used to estimate these covariances (more about the degradation in Chapter 3) and should result in inferior performance compared to techniques that make use of all available covariances. In what follows we describe a smoothing technique that uses hermitian blocks of covariance matrices from the array output matrix to decorrelate the coherent signals present in the scene. This spatial smoothing scheme, first suggested by Evans *et al.* [24 - 25] and extensively studied by Shan *et al.* [26, 32], is based on a a preprocessing scheme that partitions the total array of sensors into subarrays and then generates the average of the subarray output covariance matrices. Shan *et al.* have shown that when this average of subarray covariance matrices is used in conjunction with the multiple signal classification technique developed by Schmidt [19] in the case of independent and identical sensor noise, it is possible to estimate all directions of arrival irrespective of their degree of correlation. As shown below this method also makes use of a larger number of sensor elements than the conventional ones, and in particular requires $2K$ sensor elements to estimate K coherent directions of arrival.

Further, we also analyze an improved spatial smoothing scheme and prove that at most $[3K/2]$ elements are enough to estimate any K directions of arrival [33]. (The symbol [x] stands for the smallest integer greater than or equal to x.) In addition to the forward subarrays, this scheme makes use of complex conjugated backward subarrays of the original array to achieve superior performance. A special case of the general situation, where the multipath coefficients are treated to be real, is proved in [34]. However, this is an unrealistic assumption, as in practice all multipath coefficients will be invariably complex numbers and in that case it is necessary to reason differently.

For clarity of presentation, the next section deals with a completely coherent situation and proves that to estimate any K such directions of arrival it is sufficient to have an array of $[3K/2]$ sensors. The proof for the general situation is sketched in Appendix 2.A.

2.3.3 Spatial Smoothing Schemes: Direction Finding in a Coherent Environment

Consider a uniform linear array consisting of M identical sensors and receiving signals from K narrowband coherent sources that arrive at the array from directions $\theta_1, \theta_2, \cdots, \theta_K$. At any instant these K signals $u_1(t), u_2(t), \cdots, u_K(t)$, are phase-delayed, amplitude-weighted replicas of one of them – say, the first – and hence as in (2.68) $u_k(t) = \alpha_k u_1(t);\ k = 1, 2, \cdots, K$, where α_k represents the complex attenuation of the kth signal with respect to the first signal, $u_1(t)$. Here again the interelement distance is taken to be half wavelength, and it is assumed that the signals and noises are stationary, zero mean uncorrelated random processes, and further, the noises are assumed to be uncorrelated and identical between themselves with common variance σ^2. As before, the received signal $x_i(t)$ at the ith sensor element can be expressed as in (2.49) with $d_i = (i-1)$. Further using (2.68) in (2.54), and under the normalization $E = [\,|u_1(t)|^2] = 1$, the source covariance matrix $\mathbf{R}_u$ in (2.58) has the form

$$\mathbf{R}_u = E[\mathbf{u}(t)\mathbf{u}^\dagger(t)] = \alpha\alpha^\dagger \qquad (2.102)$$

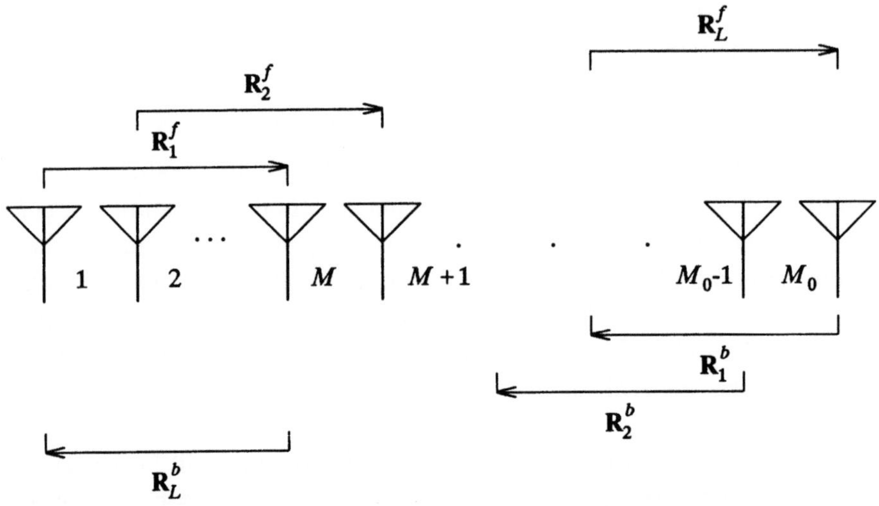

Fig. 2.13 The forward/backward spatial smoothing scheme.

where

$$\boldsymbol{\alpha} = \left[\alpha_1, \alpha_2, \cdots, \alpha_K \right]^T .$$
(2.103)

From our assumptions the array output matrix $\mathbf{R}$ in (2.56) - (2.57) can be written as

$$\mathbf{R} = \mathbf{A}\boldsymbol{\alpha}\boldsymbol{\alpha}^\dagger\mathbf{A}^\dagger + \sigma^2\mathbf{I} = \mathbf{b}\,\mathbf{b}^\dagger + \sigma^2\mathbf{I}$$
(2.104)

where $\mathbf{b}$ is as given in (2.70). If $\lambda_1 \geq \lambda_2 \geq \cdots \geq \lambda_M$ and $\beta_1, \beta_2, \cdots, \beta_M$ are the eigenvalues and the corresponding eigenvectors of $\mathbf{R}$, then reasoning as before it follows that

$$\lambda_2 = \lambda_3 = \cdots = \lambda_M = \sigma^2 ,$$
(2.105)

and hence

$$\beta_i^\dagger \mathbf{b} = 0 , \quad i = 2, 3, \cdots, M .$$
(2.106)

Because of their Vandermonde structure, no linear combination of direction vectors can result in another direction vector. Consequently $\mathbf{b}$ is no longer a legitimate direction vector and hence (2.106) will not be able to estimate any true arrival angles. The crucial role played by the nonsingularity of $\mathbf{R}_u$ in this discussion has prompted

Evans *et al.* and subsequently Shan *et al.* to introduce a preprocessing scheme [24 - 26] which guarantees full rank condition for the equivalent $\mathbf{R}_u$ in (2.102) even when the signals are all coherent. This preprocessing spatial smoothing scheme starts by dividing a uniform linear array with M_0 sensors into overlapping subarrays of size M, with sensors { 1, 2, $\cdots$, M } forming the first subarray, sensors { 2, 3, $\cdots$, $M+1$ } forming the second subarray, etc. up to the last subarray formed by sensors { M_0-M+1, M_0-M+2, $\cdots$, M_0 } (see Fig. 2.13). Let $\mathbf{x}_l^f(t)$ stand for the output of the lth subarray for l = 1, 2, $\cdots$, $L \triangleq M_0-M+1$, where L denotes the total number of these forward subarrays. Thus

$$\mathbf{x}_l^f(t) = \left[x_l(t), x_{l+1}(t), \cdots, x_{l+M-1}(t) \right]^T , \quad l = 1, 2, \cdots, L, \quad (2.107)$$

and proceeding as in (2.49) - (2.55) this can be rewritten as

$$\mathbf{x}_l^f(t) = \mathbf{A}\,\mathbf{B}^{l-1}\mathbf{u}(t) + \mathbf{n}_l(t), \quad l = 1, 2, \cdots, L, \quad (2.108)$$

where $\mathbf{B}^{l-1}$ denotes the $(l-1)$ power of the $K \times K$ diagonal matrix

$$\mathbf{B} = \begin{bmatrix} \nu_1 & & & \mathbf{O} \\ & \nu_2 & & \\ & & \ddots & \\ \mathbf{O} & & & \nu_K \end{bmatrix} \quad (2.109)$$

with

$$\nu_i = e^{-j\omega_i} , \quad \omega_i = \pi\cos\theta_i ; \quad i = 1, 2, \cdots, K . \quad (2.110)$$

The covariance matrix of the lth subarray is given by

$$\mathbf{R}_l^f = E\left[\mathbf{x}_l^f(t)\,(\mathbf{x}_l^f(t))^\dagger \right]$$

$$= \mathbf{A}\,\mathbf{B}^{l-1}\mathbf{R}_u\,(\mathbf{B}^{l-1})^\dagger\mathbf{A}^\dagger + \sigma^2\mathbf{I} . \quad (2.111)$$

Define the forward spatially smoothed covariance matrix, $\mathbf{R}^f$ as the average value of the forward subarray covariance matrices and this gives

$$\mathbf{R}^f = \frac{1}{L} \sum_{l=1}^{L} \mathbf{R}_l^f = \mathbf{A} \left[\frac{1}{L} \sum_{l=1}^{L} \mathbf{B}^{l-1} \mathbf{R}_u (\mathbf{B}^{l-1})^\dagger \right] \mathbf{A}^\dagger + \sigma^2 \mathbf{I}$$

$$= \mathbf{A} \mathbf{R}_u^f \mathbf{A}^\dagger + \sigma^2 \mathbf{I} \tag{2.112}$$

where

$$\mathbf{R}_u^f = \frac{1}{L} \sum_{l=1}^{L} \mathbf{B}^{l-1} \mathbf{R}_u (\mathbf{B}^{l-1})^\dagger. \tag{2.113}$$

In a completely coherent environment, $\mathbf{R}_u$ is given by (2.102) and thus

$$\mathbf{R}_u^f = \frac{1}{L} \left[\boldsymbol{\alpha}, \mathbf{B}\boldsymbol{\alpha}, \mathbf{B}^2\boldsymbol{\alpha}, \cdots, \mathbf{B}^{L-1}\boldsymbol{\alpha} \right] \begin{bmatrix} \boldsymbol{\alpha}^\dagger \\ (\mathbf{B}\boldsymbol{\alpha})^\dagger \\ (\mathbf{B}^2\boldsymbol{\alpha})^\dagger \\ \vdots \\ (\mathbf{B}^{L-1}\boldsymbol{\alpha})^\dagger \end{bmatrix} = \frac{1}{L} \mathbf{C}\mathbf{C}^\dagger, \tag{2.114}$$

where

$$\mathbf{C} = \left[\boldsymbol{\alpha}, \mathbf{B}\boldsymbol{\alpha}, \mathbf{B}^2\boldsymbol{\alpha}, \cdots, \mathbf{B}^{L-1}\boldsymbol{\alpha} \right]$$

$$= \begin{bmatrix} \alpha_1 & & & \mathbf{O} \\ & \alpha_2 & & \\ & & \ddots & \\ \mathbf{O} & & & \alpha_K \end{bmatrix} \begin{bmatrix} 1 & \nu_1 & \nu_1^2 & \cdots & \nu_1^{L-1} \\ 1 & \nu_2 & \nu_2^2 & \cdots & \nu_2^{L-1} \\ \vdots & \vdots & \vdots & \ddots & \vdots \\ 1 & \nu_K & \nu_K^2 & \cdots & \nu_K^{L-1} \end{bmatrix} \triangleq \mathbf{D}\mathbf{V}. \tag{2.115}$$

Clearly the rank of $\mathbf{R}_u^f$ is equal to the rank of $\mathbf{C}$. Since $\mathbf{C} = \mathbf{D}\mathbf{V}$ and the square matrix $\mathbf{D}$ is of full rank, the rank of $\mathbf{C}$ is the same as that of $\mathbf{V}$. Now the rank of the $K \times L$ Vandermonde matrix $\mathbf{V}$ is $\rho(\mathbf{V}) = \min(K,L)$ and hence $\rho(\mathbf{V}) = K$ if and only if $L \geq K$. Thus, if $L = M_0 - M + 1 \geq K$ or equivalently $M_0 \geq M + K - 1$, the smoothed signal covariance matrix $\mathbf{R}_u$ is nonsingular and $\mathbf{R}^f$ has exactly the same form as the covariance matrix for a noncoherent case. Therefore the conclusions in (2.59) - (2.64) will hold good for $\mathbf{R}^f$ in (2.112), and as pointed out by Shan $et\ al.$, one can successfully apply the eigenstructure methods to this smoothed covariance matrix regardless of the coherence of the signals. However, in this case, the number of sensor elements M_0 must be at least $(M + K - 1)$, and recalling that the size M of each subarray must also be at least $K + 1$, it follows that the minimum number of sensors needed is $2K$ compared to $M_0 = K + 1$ for the conventional one. In what follows we show that by simultaneous use of forward and appropriate backward subarrays, the number of elements required can be further reduced to $[3K/2]$.

Toward this purpose additional L backward subarrays are generated from the same set of sensors by grouping elements at $\{M_0, M_0 - 1, \cdots, M_0 - M + 1\}$ to form the first backward subarray and elements at $\{M_0 - 1, M_0 - 2, \cdots, M_0 - M\}$ to form the second one, etc. (see Fig. 2.13). Let $\mathbf{x}_l^b(t)$ denote the complex conjugate of the output of the lth backward subarray for $l = 1, 2, \cdots, L$, where L as before denotes the total number $(M_0 - M + 1)$ of these subarrays. Thus

$$\mathbf{x}_l^b(t) = \left[x_{M_0 - l + 1}^*(t), x_{M_0 - l}^*(t), \cdots, x_{L - l + 1}^*(t) \right]^T,$$

$$l = 1, 2, \cdots, L \tag{2.116}$$

and as in (2.108) this can be rewritten as

$$\mathbf{x}_l^b(t) = \mathbf{A}\left(\mathbf{B}^{M_0 - l}\, \mathbf{u}(t) \right)^* + \tilde{\mathbf{n}}_l^*(t)$$

$$= \mathbf{A}\mathbf{B}^{l-1}\left(\mathbf{B}^{M_0 - 1}\, \mathbf{u}(t) \right)^* + \tilde{\mathbf{n}}_l^*(t), \quad l = 1, 2, \cdots, L \tag{2.117}$$

where $\mathbf{B}$ is as defined in (2.109). The covariance matrix of the lth backward subarray is given by

$$\mathbf{R}_l^b = E[\mathbf{x}_l^b(t)\mathbf{x}_l^{b\dagger}(t)] = \mathbf{A}\mathbf{B}^{l-1}\mathbf{R}_{\tilde{u}}(\mathbf{B}^{l-1})^\dagger\mathbf{A}^\dagger + \sigma^2\mathbf{I} \quad (2.118)$$

with

$$\mathbf{R}_{\tilde{u}} \triangleq \mathbf{B}^{-(M_0-1)}E[\mathbf{u}^*(t)\mathbf{u}^T(t)](\mathbf{B}^{-(M_0-1)})^\dagger$$

$$= \mathbf{B}^{-(M_0-1)}\mathbf{R}_u^*(\mathbf{B}^{-(M_0-1)})^\dagger \tag{2.119}$$

and $\mathbf{R}_u$ as defined in (2.58). As before, define the spatially smoothed backward subarray covariance matrix $\mathbf{R}^b$ as the average value of these subarray covariance matrices. This gives

$$\mathbf{R}^b = \frac{1}{L}\sum_{l=1}^{L}\mathbf{R}_l^b = \mathbf{A}\left[\frac{1}{L}\sum_{l=1}^{L}\mathbf{B}^{l-1}\mathbf{R}_{\tilde{u}}(\mathbf{B}^{l-1})^\dagger\right]\mathbf{A}^\dagger + \sigma^2\mathbf{I}$$

$$= \mathbf{A}\mathbf{R}_u^b\mathbf{A}^\dagger + \sigma^2\mathbf{I}, \tag{2.120}$$

where

$$\mathbf{R}_u^b = \frac{1}{L}\sum_{l=1}^{L}\mathbf{B}^{l-1}\mathbf{R}_{\tilde{u}}(\mathbf{B}^{l-1})^\dagger. \tag{2.121}$$

In a completely coherent environment $\mathbf{R}_u$ is given by (2.102) and in that case using (2.102) in (2.119) $\mathbf{R}_{\tilde{u}}$ simplifies to

$$\mathbf{R}_{\tilde{u}} = \boldsymbol{\delta}\boldsymbol{\delta}^\dagger, \tag{2.122}$$

where

$$\boldsymbol{\delta} = \left[\delta_1, \delta_2, \cdots, \delta_K\right]^T \tag{2.123}$$

and

$$\delta_k = \alpha_k^* \nu_k^{-(M_0-1)}, \quad k = 1, 2, \cdots, K \tag{2.124}$$

with $\nu_k; k = 1, 2, \cdots, K$ as defined in (2.110). Finally, with (2.122) in (2.121) it simplifies to

$$\mathbf{R}_u^b = \frac{1}{L} \left[\boldsymbol{\delta}, \mathbf{B}\boldsymbol{\delta}, \mathbf{B}^2\boldsymbol{\delta}, \cdots, \mathbf{B}^{L-1}\boldsymbol{\delta} \right] \begin{bmatrix} \boldsymbol{\delta}^\dagger \\ (\mathbf{B}\boldsymbol{\delta})^\dagger \\ \vdots \\ (\mathbf{B}^{L-1}\boldsymbol{\delta})^\dagger \end{bmatrix} = \frac{1}{L} \mathbf{E}\,\mathbf{E}^\dagger , \quad (2.125)$$

where

$$\mathbf{E} = \left[\boldsymbol{\delta}, \mathbf{B}\boldsymbol{\delta}, \mathbf{B}^2\boldsymbol{\delta}, \cdots, \mathbf{B}^{L-1}\boldsymbol{\delta} \right]$$

$$= \begin{bmatrix} \delta_1 & & & \mathbf{O} \\ & \delta_2 & & \\ & & \ddots & \\ \mathbf{O} & & & \delta_K \end{bmatrix} \begin{bmatrix} 1 & \nu_1 & \nu_1^2 & \cdots & \nu_1^{L-1} \\ 1 & \nu_2 & \nu_2^2 & \cdots & \nu_2^{L-1} \\ \vdots & \vdots & \vdots & \ddots & \vdots \\ 1 & \nu_K & \nu_K^2 & \cdots & \nu_K^{L-1} \end{bmatrix} \triangleq \mathbf{F}\mathbf{V} \quad (2.126)$$

with $\mathbf{V}$ as given in (2.115). Reasoning as before, it is easy to see that the spatially smoothed backward subarray covariance matrix $\mathbf{R}^b$ will be of full rank so long as $\mathbf{R}_u^b$ is nonsingular, and this is guaranteed whenever $L \geq K$. Again it follows that the backward subarray averaging scheme also requires at most $2K$ sensor elements to estimate the directions of arrival of any K sources irrespective of their coherence.

It remains to show that by simultaneous use of the forward and backward subarray averaging schemes, it is possible to further reduce the number of extra sensor elements. To see this, define the forward/backward smoothed covariance matrix $\tilde{\mathbf{R}}$ as the mean of $\mathbf{R}^f$ and $\mathbf{R}^b$; i.e.

$$\tilde{\mathbf{R}} = \frac{\mathbf{R}^f + \mathbf{R}^b}{2} . \quad (2.127)$$

Using (2.112), (2.114), (2.120) and (2.125) in (2.127) we have

$$\tilde{R} = A \left[\frac{1}{2L} \left(C C^\dagger + E E^\dagger \right) \right] A^\dagger + \sigma^2 I = A \tilde{R}_u A^\dagger + \sigma^2 I , \quad (2.128)$$

where

$$\tilde{R}_u = \frac{R_u^f + R_u^b}{2} = \frac{1}{2L} [C C^\dagger + E E^\dagger] = \frac{1}{2L} G G^\dagger . \quad (2.129)$$

Here

$$G = \left[\alpha, B\alpha, B^2\alpha, \cdots, B^{L-1}\alpha, \delta, B\delta, B^2\delta, \cdots, B^{L-1}\delta \right]$$

$$= \left[DV \quad FV \right] = D \left[V \quad HV \right] = D G_0 , \quad (2.130)$$

with D, V as in (2.115) and

$$H = \begin{bmatrix} \varepsilon_1 & & & O \\ & \varepsilon_2 & & \\ & & \ddots & \\ O & & & \varepsilon_K \end{bmatrix} \quad (2.131)$$

where

$$\varepsilon_k = \delta_k / \alpha_k , \quad k = 1, 2, \cdots, K . \quad (2.132)$$

We will now prove that the modified source covariance matrix $\tilde{R}_u$ given by (2.129) is nonsingular regardless of the coherence of the K signal sources so long as $2L \geq K$, provided that whenever equality holds among some of the members of the set $\{ \varepsilon_k \}_{k=1}^{K}$ in (2.132), its largest subset with equal entries must at most be of size L.

To appreciate this restriction, first consider the case where all $\varepsilon_k; k = 1, 2, \cdots, K$ are equal. In that case it is easy to see that G_0 and hence $\tilde{R}_u$ will be of rank L irrespective of the backward smoothing process. However, in practice this equality condition almost never occurs. This is because α_k in (2.68), which represents the complex attenuation of the kth source with respect to the reference source, is a signal property, and δ_k in (2.124), which is a function of the interelement phase delay of the kth source with respect to the reference element, is mainly an array geometry property. Thus in an actual situation all $\varepsilon_k; k = 1, 2, \cdots, K$ will be distinct and the simultaneous equality condition for all of them makes it an event that almost never occurs. From these arguments it also follows that the above restrictions on the equality among some of the $\varepsilon_k s$ will almost always be satisfied. To be specific with regard to these restrictions, we will assume that

$$\varepsilon_i \neq \varepsilon_j \text{ , for } any \ i = 1, 2, \cdots, L, \text{ and } j = L+1, \cdots, K . \quad (2.133)$$

A special case of this general situation, where all $\alpha_k; k = 1, 2, \cdots, K$, in (2.68) are real, is treated in [34]. In that case using (2.124), (2.131) and (2.132) in (2.130) it is easy to see that G_0 is of Vandermonde type and hence is of rank K so long as $2L \geq K$. This, however, is an unrealistic assumption, as in practice, all $\alpha_k s$ will be invariably complex numbers, and in that case it is necessary to argue differently as follows:

From (2.129), $\tilde{R}_u$ will be nonsingular so long as G is of full row rank, and using (2.130) this is further equivalent to having full row rank for G_0. Clearly, for G (or G_0) to have full row rank it is necessary that $2L \geq K$ and with $L = M_0 - M + 1$, this reduces to $2M_0 \geq 2M + K - 2$. Again recalling that in the presence of K signals the size M of each subarray must be at least $K + 1$, it follows that the number of sensors M_0 needed must satisfy $2M_0 \geq 3K$ or, equivalently, the minimum number of sensors must be at least $[3K/2]$. To see that this is also sufficient, consider the quadratic product

$$\mathbf{y}^\dagger G_0 G_0^\dagger \mathbf{y} = \mathbf{y}^\dagger V V^\dagger \mathbf{y} + \mathbf{y}^\dagger H V V^\dagger H^\dagger \mathbf{y} \quad (2.134)$$

where $\mathbf{y}$ is any arbitrary $K \times 1$ vector. We will show that

$$\mathbf{y}^\dagger G_0 G_0^\dagger \mathbf{y} > 0 \quad (2.135)$$

for any $\mathbf{y} \neq 0$, thus proving the positive definite property of $G_0 G_0^\dagger$ or $\tilde{R}_u$. Clearly (2.135) needs to be demonstrated only for a typical $\mathbf{y}_0 \in N(V^\dagger)$, the null space of $V^\dagger$. In that case $V^\dagger \mathbf{y}_0 = 0$ and hence the first term in (2.134) also reduces to zero. To prove our claim it is enough to show that for such a typical $\mathbf{y}_0$, the vector $H^\dagger \mathbf{y}_0$ does not belong to $N(\mathbf{V}^\dagger)$. Since the Vandermonde structured matrix $\mathbf{V}^\dagger$ is of full row rank L, the dimension of $N(\mathbf{V}^\dagger)$ is $K-L$. Let $\mathbf{v}_{L+1}$, $\mathbf{v}_{L+2}$, $\cdots$, $\mathbf{v}_K$ be a set of linearly independent basis vectors for $N(\mathbf{V}^\dagger)$. With respect to the basis vectors for the K-dimensional space, these $K \times 1$ null space basis vectors can always be selected such that [35],

$$\mathbf{v}_l = \left[v_{1l}, v_{2l}, \cdots, v_{Ll}, 0, \cdots, 0, 1, 0, \cdots, 0 \right]^T . \qquad (2.136)$$

(In (2.136) the 1 is at the lth location.) These $\mathbf{v}_l$, $l = L+1, L+2$, $\cdots$, K are linearly independent and, moreover, for any $j \in \{ L+1, L+2, \cdots, K \}$, using the diagonal nature of $\mathbf{H}$ it is also easy to see that $\mathbf{H}^\dagger \mathbf{v}_j$ is linearly independent of the remaining $\mathbf{v}_l$, $l = L+1$, $\cdots, K$, $l \neq j$. Further the pair $\mathbf{v}_j$ and $\mathbf{H}^\dagger \mathbf{v}_j$, $j = L+1, L+2, \cdots, K$, is also linearly independent of each other. To see this note that because of the full row rank property of $\mathbf{V}^\dagger$, at least one of the v_{il}; $i = 1, 2, \cdots, L$ in (2.136) must be nonzero for every l. Let $v_{i_o j}$ be such an entry in $\mathbf{v}_j$. Then the minor formed by the i_o^{th} and j^{th} rows of the matrix $[\mathbf{v}_j \,|\, \mathbf{H}^\dagger \mathbf{v}_j]$ has the form

$$\begin{vmatrix} v_{i_o j} & \varepsilon_{i_o}^* v_{i_o j} \\ 1 & \varepsilon_j^* \end{vmatrix} = v_{i_o j} (\varepsilon_j^* - \varepsilon_{i_o}^*) \qquad (2.137)$$

and is nonzero from (2.133). Thus the matrix $[\mathbf{v}_j \,|\, \mathbf{H}^\dagger \mathbf{v}_j]$ is of rank 2. This proves the linear independence of $\mathbf{v}_j$ and $\mathbf{H}^\dagger \mathbf{v}_j$. From the above discussion it follows that $\mathbf{H}^\dagger \mathbf{v}_j$ is linearly independent of $\mathbf{v}_j$, $j = L+1$, $L+2, \cdots, K$ and hence $\mathbf{H}^\dagger \mathbf{v}_j \notin N(\mathbf{V}^\dagger)$, $j = L+1, L+2, \cdots, K$. Now for any $\mathbf{y}_0 \in N(\mathbf{V}^\dagger)$ we have

$$\mathbf{y}_0 = \sum_{j=L+1}^{K} k_j \mathbf{v}_j , \qquad (2.138)$$

which gives

$$\mathbf{H}^\dagger \mathbf{y}_0 = \sum_{j=L+1}^{K} k_j \mathbf{H}^\dagger \mathbf{v}_j . \qquad (2.139)$$

Since $\mathbf{H}^\dagger \mathbf{v}_j$, $j = l+1, L+2, \cdots, K$ are linearly independent and all k_j cannot be zero in (2.139), it follows that $\mathbf{H}^\dagger \mathbf{y}_0 \notin N(\mathbf{V}^\dagger)$ and hence

$\mathbf{V}^\dagger \mathbf{H}^\dagger \mathbf{y}_0 \neq 0$. This proves our claim and establishes that $\mathbf{G}_0$ will be non-singular under the mild restrictions in (2.133). Within these conditions $\tilde{\mathbf{R}}_u$ is also of rank K and hence the eigenvalues of $\tilde{\mathbf{R}}$ satisfy $\tilde{\lambda}_1 \geq \tilde{\lambda}_2 \geq \cdots \geq \tilde{\lambda}_K > \tilde{\lambda}_{K+1} = \tilde{\lambda}_{K+2} = \cdots = \sigma^2$. Consequently, as in (2.67) the nulls of $\tilde{Q}(\omega)$ given by

$$\tilde{Q}(\omega) = \sum_{k=K+1}^{M} | \tilde{\beta}_k^\dagger \mathbf{a}(\omega) |^2 = 1 - \sum_{k=1}^{K} | \tilde{\beta}_k^\dagger \mathbf{a}(\omega) |^2 \qquad (2.140)$$

or the peaks of

$$\tilde{P}(\omega) = \frac{1}{\tilde{Q}(\omega)}$$

correspond to the actual directions of arrival. Here $\tilde{\beta}_1, \tilde{\beta}_2, \cdots, \tilde{\beta}_M$ are the eigenvectors of $\tilde{\mathbf{R}}$ corresponding to the eigenvalues $\tilde{\lambda}_1, \tilde{\lambda}_2, \cdots, \tilde{\lambda}_M$.

To summarize, we have shown that so long as the number of sensor elements is at least $[3K/2]$, (with K representing the number of signal sources present in the scene), it is almost always possible to estimate all arrival angles irrespective of the signal correlations, by simultaneous use of the forward and backward subarray averaging scheme. Since the smoothed covariance matrix $\tilde{\mathbf{R}}$ in (2.127) has exactly the same form as the covariance matrix for some noncoherent case as in (2.57) - (2.59), the eigenstructure-based techniques can be applied to this smoothed covariance matrix irrespective of the coherence of the signals, to estimate their directions of arrival successfully.

Appendix 2.A extends the proof for this forward/backward smoothing scheme to a mixed source scene consisting of partially correlated signals with complete coherence among some of them.

Fig. 2.14 represents a coherent source scene where the reference signal arriving along 70° undergoes multipath reflection resulting in three additional coherent arrivals along 42°, 110° and 127°. A six-element uniform array is used to receive these signals. The input signal-to-noise ratio of the reference signal is 10 dB, and the attenuation coefficients of the three coherent sources are taken to be $0.4 + j0.8$, $-0.3 - j0.7$ and $0.5 - j0.6$, respectively. Five hundred data samples are used to estimate the array output covariance matrix. The application of the conventional eigenstructure method [19] to this

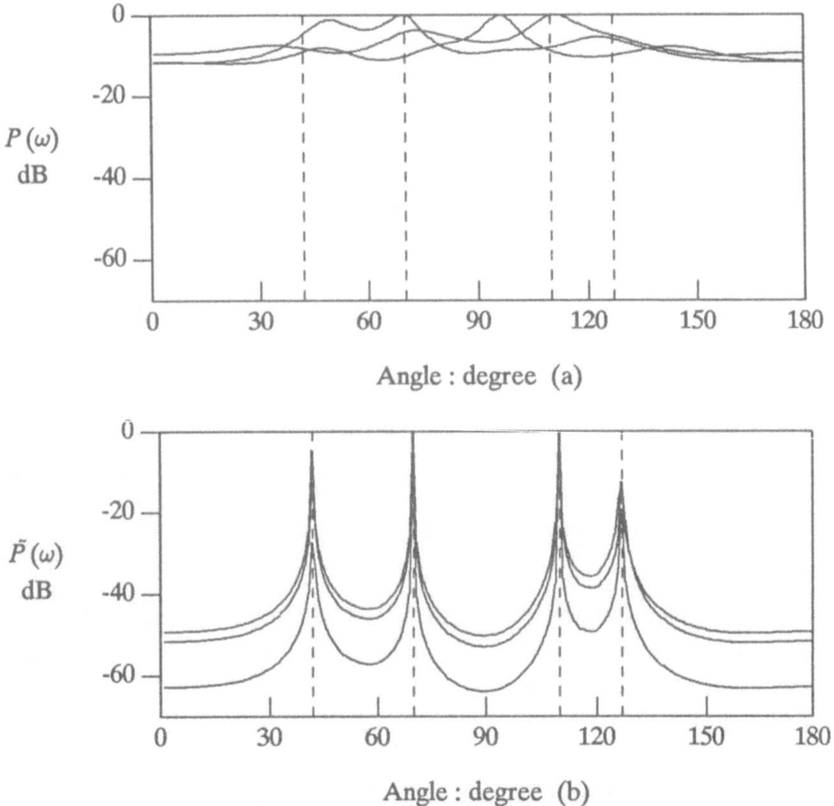

Fig. 2.14 Direction Finding in a coherent scenario. (a) $P(\omega)$ using the conventional MUSIC algorithm. (b) $\tilde{P}(\omega)$ using the forward/backward smoothing scheme.

covariance matrix results in Fig. 2.14.a. However, first applying the forward/backward spatial smoothing scheme with two forward and two backward ($L = 2$) subarrays of five ($M = 5$) sensors each, and then reapplying the eigenstructure technique on the smoothed covariance matrix $\tilde{\mathbf{R}}$ results in Fig. 2.14.b. All four directions of arrival can be clearly identified and the improvement in performance in terms of resolvability, irrespective of the signal coherence, is also visible in this case.

Thus, by simultaneous use of a set of forward and complex conjugated backward subarrays, it is possible to reduce the extra sensor elements required to create a smoothed array output covariance matrix to one half the number of the coherent signals present in the scene. Further, this smoothed covariance matrix is shown to be structurally identical to that present in a noncoherent situation, thus enabling one to correctly identify all directions of arrival by incorporating the eigenstructure-based techniques [19] on this smoothed matrix. This is a considerable saving compared to the forward-only smoothing scheme [24 - 26] which requires as many extra sensor elements as the total number of coherent signals.

So far we have assumed that an ensemble average of the array output covariances are available. When these ensemble averages are not available, they have to be estimated from the array output data. Generally, a finite data sample is used and estimation is carried out for the unknowns of interest using the maximum likelihood procedure [36]. In the next chapter we study the statistical properties of these estimated smoothed covariance matrices and their associated sample estimators for direction finding.

2.4 Augmentation and Other Techniques

Having considered the basic high resolution techniques for direction finding in some detail, it is now appropriate to discuss some of the closely related methods that operate under different considerations. In general, array performance can be optimized with respect to improvements in detection and resolution of closely spaced sources by making use of specific signal properties such as their crosscorrelation, polarization or mean values. We will first describe a technique that, for uncorrelated signals, exploits the array geometry to improve upon its detection capacity.

2.4.1 Augmentation Technique

A recurring question in array design for both signal reception and parameter estimation is that of how to deploy the elements of a sparse array beneficially [5]. In the present context of direction finding, this is also significant due to considerations on improvements in performance in terms of detecting and resolving a larger number of sources beyond the conventional limits by suitable modifications of the array geometry.

To investigate this further, consider the source scene described in section 2.3 and assume the K sources in (2.49) to be uncorrelated with each other. This gives the correlation r_{ij} between the ith and j^{th} sensor elements to be

$$r_{ij} = E[x_i(t)x_j^*(t)] = \sum_{k=1}^{K} P_k e^{-j(d_i-d_j)\cos\theta_k} + \sigma^2 \delta_{ij} ,$$

$$i,j = 1, 2, \cdots, M . \tag{2.141}$$

Here $P_k, k = 1, 2, \cdots, K$ represents the signal power associated with the kth source.

A theorem by Carathéodory in the context of finite moment problems is of special significance here. It states that given $N+1$ complex constants $c_0, c_1, \cdots, c_N$ not all zero, $N > 1$, with $c_{-n}^* = c_n$, then there exists an integer K, $1 \le K \le N$ and certain constants $\alpha_k > 0$ and ω_k, for $k = 1, 2, \cdots, K$ such that

$$c_m = \sum_{k=1}^{K} \alpha_k e^{jm\omega_k} + \alpha_0 \delta_m , \quad m = 1, 2, \cdots, N . \tag{2.142}$$

Further, the integer K and constants α_k and ω_k are determined uniquely [3] [37]. Comparison of (2.141) and (2.142) reveals that the autocorrelation lags represented by (2.141) have a natural Carathéodory representation as given in (2.142). Further, the analogy is exact if the M array element locations are spatially distributed in such a way that the set of integers $\{m\}$ implied by the set of differences $d_j - d_i = m, j \ge i, i,j = 1, 2, \cdots, M$ represents every integer in the set $\{ 0, 1, 2, \cdots, N \}$ where $N \le M(M\text{-}1)/2$. Then with M elements we have $(N+1)$ autocorrelation lags $r(m)$, where

$$r(m) = r(j-i) = \sum_{k=1}^{K} P_k e^{j\pi m \omega_k} + \sigma^2 \delta_m , \quad m = 1, 2, \cdots, N .\tag{2.143}$$

For a given M, such an element location set $d_1, d_2, \cdots, d_M$ (normalized to $\lambda/2$), forms a sequence, which is defined to be a Carathéodory

____

(3) In general α_0 can be positive (including zero) or negative depending on whether the hermitian Toeplitz matrix generated by the original sequence $c_0, c_1, \cdots, c_N$ is nonnegative definite or otherwise. For an autocorrelation sequence, $\alpha_0 \ge 0$.

sequence (C-sequence) of length M [42]. (For example, a four-element C-sequence is { 0, 2, 5, 6 }.)

Notice that (2.143) is functionally identical to (2.84) and hence to compute the actual directions of arrival once again we can resort to the technique described in section 2.3.2. In that case the hermitian Toeplitz matrix $\mathbf{R}_a$ generated by $r(0)$, $r(1)$, $\cdots$, $r(N)$ will be identically the same as the output covariance matrix of an array with $N+1$ uniformly placed elements. We refer to $\mathbf{R}_a$ as the augmented covariance matrix, which is derived from the smaller covariance matrix $\mathbf{R}$ of the M-element array. Utilizing $\mathbf{R}_a$, one expects therefore to be able to handle a greater number of signal arrivals than could be handled using $\mathbf{R}$. In fact using (2.143) and the above definition, $\mathbf{R}_a$ can be rewritten as

$$\mathbf{R}_a = \mathbf{A}_0 \mathbf{R}_u \mathbf{A}_0^\dagger + \sigma^2 \mathbf{I}_N \qquad (2.144)$$

where $\mathbf{A}_0$ is an $N \times K$ matrix given by

$$\mathbf{A}_0 = \sqrt{M} \left[\mathbf{a}_0(\omega_1), \mathbf{a}_0(\omega_2), \cdots, \mathbf{a}_0(\omega_K) \right]$$

with $\mathbf{a}_0(\omega_k)$ representing the normalized direction vector corresponding to a uniform array with N-elements (refer to (2.52)) and

$$\mathbf{R}_u = diag[P_1, P_2, \cdots, P_K].$$

Arguing as before, the N-K eigenvectors of $\mathbf{R}_a$ corresponding to its lowest eigenvalue σ^2, are orthogonal to the "augmented" direction vectors $\mathbf{a}_0(\omega_k)$, $k = 1, 2, \cdots, K$ and utilizing them in an expression similar to (2.66), the actual arrival angles can be evaluated so long as $K \leq N$. Thus the maximum number of sources whose directions of arrival can be determined using the above augmented matrix is equal to N. Considering that a conventional array with M elements can only estimate at most $(M-1)$ directions of arrival, this is rather remarkable since $N > M$ and in principle could be as high as $M(M-1)/2$.

In a typical design one will be interested in studying the optimum distribution of the array elements. For a given M, let N_M denote the maximum attainable value for N over all possible sensor element placement. Then $N_M = M(M-1)/2$ if and only if $M \leq 4$. The exact

Table 2.1 Interelement spacing for optimal restricted difference basis (M = number of sensors, N_M = array length).

M	N_M	Interelement Spacing	$\dfrac{M^2}{N_M}$
3	3	.1.2.	3.0
4	6	.1.3.2.	2.667
5	9	.1.3.3.2.	2.778
6	13	.1.1.4.4.3.	2.769
7	17	.1.1.4.4.4.3.	2.882
8	23	.1.3.6.6.2.3.2.	2.783
9	29	.1.4.4.7.7.3.2.1.	2.793
10	36	.1.4.4.7.7.7.3.2.1.	2.778
11	43	.1.4.4.7.7.7.7.3.2.1.	2.814
12	50	.1.4.4.7.7.7.7.7.3.2.1.	2.88
13	58	.1.4.3.4.9.9.9.9.5.1.2.2.	2.914
14	68	.1.1.6.6.6.11.11.11.5.5.3.1.1.	2.882
15	79	.1.1.6.6.6.11.11.11.11.5.5.3.1.1.	2.848
16	90	.1.1.6.6.6.11.11.11.11.11.5.5.3.1.1.	2.844
17	101	.1.1.6.6.6.11.11.11.11.11.11.5.5.3.1.1.	2.861

values of N_M are unknown except for small Ms. Table 2.1 lists some of the optimal sequences obtained by exhaustive search. However, as summarized below, considerable effort seem to have been spent in a variety of contexts [38 - 42], for finding these optimal or near-optimal sequences.

More on Array Geometry

A closely related problem called "the representation of 1, 2, $\cdots$, n by differences" seems to have been of interest to mathematicians for a long time [38 - 41]. Rédei and Rényi [38] called the set of integers $d_1, d_2, \cdots, d_m$, a difference basis with respect to n if every positive integer k such that $0 < k \leq n$ can be represented in the form $k = d_j - d_i$. Let $m(n)$ denote the minimum value of m for given n. Rédei and Rényi proved that

$$(i) \quad \lim_{n \to \infty} \frac{m^2(n)}{n} \text{ exists,}$$

$$(ii) \quad \lim_{n \to \infty} \frac{m^2(n)}{n} = g.l.b. \frac{m^2(n)}{n} \quad (g.l.b. \text{ denotes greatest}$$
lower bound),

$$(iii) \ 2.424\cdots \le \lim_{n \to \infty} \frac{m^2(n)}{n} \le \frac{8}{3} = 2.666\cdots . \qquad (2.145)$$

Prior to this, Brauer [39] considered the problem [4] of finding a "restricted" difference basis with respect to N such that $0 = d_1 < d_2 < \cdots < d_M = N$. Notice that the minimum redundancy location sets of interest to us here are the same as Brauer's restricted difference sets. As before, let $M(N)$ denote the minimum value of M for given N. Erdös and Gál in [40] stated that for these difference sets the sequence of numbers $M^2(N)/N$ converge to a limit and moreover the results (ii) and (iii) stated above for (unrestricted) difference sets are also true for these restricted sets. However, as pointed out by Leech [41], these claims are improper because of certain incorrect substitutions in [40]. In addition to improving the original bounds in (iii) to $2.434\cdots \le \frac{m^2(n)}{n} \le 2.6646\cdots$, Leech [41] also corrected and extended the work of Erdös and Gál and showed that in the case of restricted difference sets

$$(iv) \quad \lim_{N \to \infty} \frac{M^2(N)}{N} \text{ exists,}$$

$$(v) \quad \lim_{N \to \infty} \frac{M^2(N)}{N} = g.l.b. \frac{(M(N)+\lambda)^2}{N+1} \text{ for } \lambda \ge 2,$$

$$(vi) \ 2.434\cdots \le \lim_{N \to \infty} \frac{M^2(N)}{N} \le \frac{375}{112} = 3.348\cdots . \qquad (2.146)$$

Here (iv) and (v) have been obtained by correcting Erdös and Gál's work and the left side of (vi) is an immediate consequence of the

---

(4) Brauer was primarily interested in tapping a 30 Ω rheostat using the least number of contact points such that each integral resistance from one to 30 Ω could be obtained by connecting two of the contact points in the rheostat. He proved that this problem required at least 10 contact points, i.e., $M(30) = 10$.

improved lower bound in (iii). Further, the right side is a new estimate obtained from a specific restricted difference basis generated by a new method. However for a given M, this method does not seem capable of generating restricted difference basis that at least achieves the upper bound claimed in (vi). Recently, Pearson *et al.* [42] have given constructive procedures that create superior restricted difference basis and have shown that for any $M > 3$, it is always possible to choose a restricted difference basis with respect to N such that $M^2/N < 3$. To investigate this further, let $0 = d_1 < d_2 < \cdots < d_M = N$ denote the set of distinct integers that represent a restricted difference basis for N. Define $a_i = d_{i+1} - d_i$, $i = 1, 2, \cdots, M\text{-}1$. Then $N = \sum_{i=1}^{M-1} a_i$ and for $k = 1, 2, \cdots$, let

$$
a_i = \begin{cases} 1 & 1 \leq i \leq k \\ k+2 & k+1 \leq i \leq 2k \\ k+1 & i = 2k+1. \end{cases} \qquad (2.147)
$$

Thus we have

$$
\bullet \ 1 \ \bullet \ 1 \ \bullet \ \cdots \ \bullet \ 1 \ \bullet \ k+2 \ \bullet \ k+2 \ \bullet \ \cdots \ \bullet \ k+2 \ \bullet \ k+1 \ \bullet
$$

$$
\longleftarrow \text{------} k \text{------} \longrightarrow \qquad \longleftarrow \text{--------} k \text{--------} \longrightarrow
$$

which gives $M = 2k+2$ and $N = k^2+4k+1$. It is quite easy to verify that all missing integers less than N can in fact be expressed as the difference of two elements (or equivalently as sums of consecutive a_is) in this set. Though for $k = 1$ and 2, this generates the optimal sequence, notice that $\dfrac{M^2}{N} = 4\left[1 - \dfrac{2k}{k^2+4k+1}\right] < 4$, is inferior to the upper bound in (vi), and hence generally (2.147) represents a suboptimal basis.

Clearly, the sequence a_i, $i = 1, 2, \cdots, M\text{-}1$ of distances between adjacent elements represents a restricted difference set if and only if for every positive integer $s \leq N$ there is a consecutive subsequence of the a's $a_j, a_{j+1}, \cdots, a_k$ such that $\sum_{i=j}^{k} a_i = s$. The following

Table 2.2

M	a_i	N	M^2/N
4	. 1 . 3 . 2 .	6	2.666
5	. 1 . 3 . 3 . 2 .	9	2.777
6	. 1 . 1 . 4 . 4 . 3 .	13	2.76
7	. 1 . 1 . 4 . 4 . 4 . 3 .	17	2.88

lemma gives such a restricted difference set.

Lemma

Given any integers p and n greater than *or equal to* zero, let $q = 2p + 1$ and $r = 2q + 1$. We construct a difference set with

$$a_i = \begin{cases} 1 & 1 \le i \le p \\ q+1 & p+1 \le i \le q \\ r & q+1 \le i \le q+n \\ q & q+n+1 \le i \le q+n+p \\ p+1 & i = 3p+n+2 \\ 1 & 3p+n+3 \le i \le 2q+n . \end{cases} \qquad (2.148)$$

Then it is possible to express any integer s, $0 < s \le N = (p+n+1)(4p+3)+p$ as the sum of consecutive a_i's in (2.148). Here the basis set contains $M = 4p+n+3$ elements.

$$\bullet \; 1 \cdots 1 \bullet q+1 \cdots q+1 \bullet r \cdots r \bullet q \cdots q \bullet p+1 \bullet 1 \cdots 1 \bullet$$
$$\leftarrow p \rightarrow \qquad \leftarrow p+1 \rightarrow \qquad \leftarrow n \rightarrow \qquad \leftarrow p \rightarrow \qquad \qquad \leftarrow p \rightarrow$$

The proof consists of verifying that, each integer s, $0 < s \le N$, is representable as the sum of consecutive elements of the above sequence [42]. It is also easy to show that the value of n that maximizes the ratio M^2/N is $2p+1$. This value is chosen to establish the desired result in the following theorem.

Theorem : For any given $M > 3$ [(5)], it is always possible to choose a set of integers $\mathbf{D} = \{ d_1, d_2, \cdots, d_M \}$ such that,

a) $0 = d_1 < d_2 < \cdots < d_M = N$,

b) any integer i, $0 \le i \le N$ is expressible as the difference of two elements in $\mathbf{D}$

and

c) $M^2/N < 3$.

Proof : First, for $M = 4, 5, 6, 7$, Table 2.2 presents actual sequences satisfying the above requirements. Notice that these are specific cases of (2.147) with $k = 1$ and $k = 2$. For $M \ge 8$ the proof proceeds in two parts. With $n = 2p+1$ in the above lemma, we first establish the claim for integers of the form $M = 6p+4, p = 1, 2, 3, \cdots$ and then show that the remaining cases can also be taken care of so as to satisfy (a) - (c). From the lemma for such an M, the set $\mathbf{D} = \{ d_1, d_2, \cdots, d_M \}$ generated by (2.148) satisfies (a) and (b). Moreover $M \equiv M_p = 6p+4$ and $N \equiv N_p = 6(2p^2+3p+1)$. Thus in this case

$$\frac{M^2}{N} = \frac{M_p^2}{N_p} = 3\left[1 - \frac{3p+1}{9(2p^2+3p+1)} \right] < 3 \quad \text{for } p \ge 1 , \text{(2.149)}$$

proving our claim for integers of the form $M = 6p+4, p = 1, 2, 3, \cdots$. To complete the proof we need to establish these results for $M = 8, 9$ and all $M_p < M < M_{p+1}$, with $p = 1, 2, \cdots$. Since $M_{p+1} - M_p = 6$, every such M can be expressed as $M = M_p + w$ where $|w| \le 3$, $p = 1, 2, \cdots$. If w is positive (negative), we will insert (delete) these extra elements into (from) the center portion of (2.148). In that case

---

(5) For $M = 3$, the set $d_1 = 0, d_2 = 1, d_3 = 3$ is optimal, and in this case $M^2/N = 3$.

$$
a_i = \begin{cases}
1 & 1 \le i \le p \\[6pt]
q+1 & p+1 \le i \le q \\[6pt]
r & q+1 \le i \le 2q+w \\[6pt]
q & 2q+w+1 \le i \le 2q+w+p \\[6pt]
p+1 & i = 2q+w+p+1 \\[6pt]
1 & 2q+w+p+1 < i \le 3q+w .
\end{cases}
\tag{2.150}
$$

Then $M = M_p + w = 6p+4+w$, $N = N_p + rw = 12p^2 + (18+4w)p + 3w + 6$. Since the side patterns in (2.148) have not been altered, from the lemma it is easy to see that all integers up to the above N can be expressed as the difference of two elements in **D** generated by (2.150). Further

$$
\frac{M^2}{N} = \frac{(M_p+w)^2}{N_p+(4p+3)w}
$$

$$
= 3\left[1 - \frac{6p+w+2-w^2}{3[12p^2+2(9+2w)p+3(2+w)]} \right].
\tag{2.151}
$$

Clearly the above ratio is strictly less than 3 for $p = 1$ and $-2 < w \le 3$, and for all $p \ge 2$ and $|w| \le 3$. This includes $M = 8, 9, \cdots$, and the proof is complete. Compared to other techniques for constructing restricted and unrestricted difference basis [38 - 41], this approach has two distinct advantages:

First, if we consider only *restricted* difference basis, this technique produces significantly better sets. *Unrestricted* sets, for given N, can be somewhat smaller than the restricted sets produced. On the other hand, the difference between the smallest and the largest element values (array span) in an unrestricted set can be somewhat larger. For antenna arrays, for example, an unrestricted set would use fewer antennas but would cover a larger distance than its restricted counterpart. The difference is not very great − using Rédei and Rényi's construction, an unrestricted set would require $\sqrt{2.67/3}$ times as many elements as the restricted set in (2.150) for the same N, or about 6 percent fewer elements, and would require about 17 percent more space.

In fact *no* difference set, restricted or unrestricted, can be made with even 10 percent fewer elements than one of the sets given here (for the same N). This follows from the lower bound of 2.434 in Leech [41]. It is very likely that this lower bound can be significantly increased for the restricted difference sets, by using the additional constraints they imply.

Second, the set of integers in this case is extremely easy to construct, requiring neither complex algorithms nor prohibitive computation time. The methods used by Erdös and Gál, Rédei and Rényi, and Leech [38, 40 - 41], while not computationally intractable, require complex algorithms involving construction of Galois fields of specific sizes, finding primitive roots, and modular arithmetic on polynomials. In addition, their methods are not readily adaptable to arbitrary size problems. Leech's best construction for unrestricted sets, for example, produces only sets of size $132\,(p^n + 1)$ where p is a prime and n a positive integer.

In principle, the optimal difference set of a given size, either restricted or unrestricted, can be constructed by a simple algorithm. Such an algorithm, however, requires running time exponential in the size of the array. Though this constitutes a constructive solution, a near-optimal solution such as the above that is easy to construct is of greater practical value.

2.4.2 ESPRIT, TLS-ESPRIT and GEESE

These relatively new schemes for angle-of-arrival estimation depart from the eigenvector-based high resolution methods described so far, on several important accounts. They utilize an underlying rotational invariance among signal subspaces induced by subsets of an array of sensors, to accomplish the main task of estimating the directions of arrival of incoming planar wavefronts. In an uncorrelated and identical sensor noise situation, all of them can be applied to a uniformly placed array or a pairwise matched arbitrary array with codirectional sensor doublets. Since functionally these two arrays generate the same structured data with respect to the methods under discussion, we will assume a uniform array to describe these new techniques.

ESPRIT

ESPRIT stands for Estimation of Signal Parameters via Rotational Invariance Techniques [43, 44] and the invariance among subspaces are realized here by constructing the auto- and cross- covariance matrices. With $\mathbf{x}_i^f(t)$ as in (2.107), let

$$\mathbf{x}(t) = \mathbf{x}_1^f(t) = \mathbf{A}\,\mathbf{u}(t) + \mathbf{n}_1(t) \qquad (2.152)$$

and

$$\mathbf{y}(t) = \mathbf{x}_2^f(t) = \mathbf{A}\,\mathbf{B}\,\mathbf{u}(t) + \mathbf{n}_2(t) \qquad (2.153)$$

where $\mathbf{B}$ is given by (2.109). Thus the auto- and cross-covariance matrices are given by

$$\mathbf{R}_{xx} = E[\mathbf{x}(t)\mathbf{x}^\dagger(t)] = \mathbf{A}\,\mathbf{R}_u\,\mathbf{A}^\dagger + \sigma^2\mathbf{I} \qquad (2.154)$$

and

$$\mathbf{R}_{xy} = E[\mathbf{x}(t)\mathbf{y}^\dagger(t)] = \mathbf{A}\,\mathbf{R}_u\,\mathbf{B}^\dagger\mathbf{A}^\dagger + \sigma^2\tilde{\mathbf{J}}_1 \qquad (2.155)$$

where $\tilde{\mathbf{J}}_1$ is an $M \times M$ matrix with ones along the first lower diagonal off the major diagonal and zeros elsewhere. Notice that under uncorrelated noise assumption, $\mathbf{R}_{xy} = \mathbf{A}\,\mathbf{R}_u\,\mathbf{B}^\dagger\mathbf{A}^\dagger$ for pairwise matched array with doublets. In the absence of coherent signals, σ^2 can be obtained from (2.154) by standard eigenstructure arguments [19] and eliminating the σ^2 terms from (2.154) and (2.155) we obtain

$$\mathbf{C}_{xx} = \mathbf{R}_{xx} - \sigma^2\mathbf{I} = \mathbf{A}\,\mathbf{R}_u\,\mathbf{A}^\dagger; \quad \mathbf{C}_{xy} = \mathbf{R}_{xy} - \sigma^2\tilde{\mathbf{J}}_1 = \mathbf{A}\,\mathbf{R}_u\,\mathbf{B}^\dagger\mathbf{A}^\dagger \qquad (2.156)$$

which gives

$$\mathbf{C}_{xx} - \gamma\mathbf{C}_{xy} = \mathbf{A}\,\mathbf{R}_u\,(\mathbf{I}_K - \gamma\mathbf{B}^\dagger)\mathbf{A}^\dagger. \qquad (2.157)$$

Since $\mathbf{A}$ and $\mathbf{R}_u$ are both of rank K, the singular values of the above matrix pencil are given by the roots of

$$|\mathbf{I}_K - \gamma\mathbf{B}^\dagger| = 0$$

and using (2.110) the desired singular values are

$$\gamma_k = (\mathbf{B})_{kk} = \nu_k = e^{-j\omega_k}, \quad k = 1, 2, \cdots, K. \qquad (2.158)$$

Thus, the directions of arrival can be obtained without involving a search technique, and in that respect computation and storage costs are reduced considerably.

Notwithstanding these merits, when estimates of the interelement covariances are used in these computations, subtracting the estimated noise variance from the auto- and cross-covariance matrices can at times be critical and may result in overall inferior results. To circumvent this difficulty to some extend, the TLS-ESPRIT (Total Least Squares ESPRIT) scheme processes $\mathbf{x}(t)$ and $\mathbf{y}(t)$ simultaneously [45].

TLS-ESPRIT

With $\mathbf{x}(t)$ and $\mathbf{y}(t)$ as in (2.152) - (2.153) let

$$\mathbf{z}(t) = \begin{bmatrix} \mathbf{x}(t) \\ \mathbf{y}(t) \end{bmatrix} = \begin{bmatrix} \mathbf{A} \\ \mathbf{A}\mathbf{B} \end{bmatrix} \mathbf{u}(t) + \begin{bmatrix} \mathbf{n}_1(t) \\ \mathbf{n}_2(t) \end{bmatrix} = \tilde{\mathbf{A}}\,\mathbf{u}(t) + \tilde{\mathbf{n}}(t).$$

Clearly

$$\mathbf{R}_{zz} = E[\mathbf{z}(t)\mathbf{z}^{\dagger}(t)] = \tilde{\mathbf{A}}\mathbf{R}_u\,\tilde{\mathbf{A}}^{\dagger} + \sigma^2 \Sigma_{\tilde{n}} \qquad (2.159)$$

where

$$\Sigma_{\tilde{n}} = \begin{bmatrix} \mathbf{I}_M & \tilde{\mathbf{J}}_1 \\ \\ \tilde{\mathbf{J}}_1^{\dagger} & \mathbf{I}_M \end{bmatrix}.$$

Since $\tilde{\mathbf{A}}\mathbf{R}_u\,\tilde{\mathbf{A}}^{\dagger}$ is of rank K, the generalized eigenvalues of $\mathbf{R}_{zz}$ with respect to the known matrix $\Sigma_{\tilde{n}}$ can be represented as $\lambda_1 \geq \lambda_2 \geq \cdots \geq \lambda_K > \lambda_{K+1} = \cdots = \lambda_{2M} = \sigma^2$. Further, the associated eigenvectors $\mathbf{e}_i$ satisfy the identity

$$\mathbf{R}_{zz}\,\mathbf{e}_i = \lambda_i\,\Sigma_{\tilde{n}}\,\mathbf{e}_i$$

or

$$\tilde{\mathbf{a}}^{\dagger}(\omega_k)\,\mathbf{e}_i = 0, \quad k = 1, 2, \cdots, K, \quad i = K+1, K+2, \cdots, 2M$$

where $\tilde{\mathbf{a}}(\omega_k)$, $k = 1, 2, \cdots, K$ are the K column vectors of $\tilde{\mathbf{A}}$. Equivalently, $\mathbf{e}_1, \mathbf{e}_2, \cdots, \mathbf{e}_K$ span the same subspace spanned by the column vectors of $\tilde{\mathbf{A}}$, i.e.,

$$\left[\mathbf{e}_1, \mathbf{e}_2, \cdots, \mathbf{e}_K \right] = \tilde{\mathbf{A}} \mathbf{C} \tag{2.160}$$

where $\mathbf{C}$ is some nonsingular $K \times K$ matrix. Define two $M \times K$ matrices $\mathbf{E}_x$ and $\mathbf{E}_y$ by partitioning (2.160) in an obvious manner. Thus

$$\left[\mathbf{e}_1, \mathbf{e}_2, \cdots, \mathbf{e}_K \right] = \begin{bmatrix} \mathbf{E}_x \\ \mathbf{E}_y \end{bmatrix}.$$

Then

$$\mathbf{E}_x = \mathbf{A}\mathbf{C}, \quad \mathbf{E}_y = \mathbf{A}\mathbf{B}\mathbf{C} \tag{2.161}$$

and

$$\left[\mathbf{E}_x \quad \mathbf{E}_y \right] = \mathbf{A} \left[\mathbf{C} \quad \mathbf{B}\mathbf{C} \right] \tag{2.162}$$

which gives

$$\mathbf{E}_{xy} \triangleq \begin{bmatrix} \mathbf{E}_x^\dagger \\ \mathbf{E}_y^\dagger \end{bmatrix} \left[\mathbf{E}_x \quad \mathbf{E}_y \right] = \begin{bmatrix} \mathbf{C}^\dagger \\ \mathbf{C}^\dagger \mathbf{B}^\dagger \end{bmatrix} \mathbf{A}^\dagger \mathbf{A} \left[\mathbf{C} \quad \mathbf{B}\mathbf{C} \right]. \tag{2.163}$$

$\mathbf{E}_{xy}$ is nonnegative definite hermitian and is of rank K. Thus $\mathbf{E}_{xy}$ has the representation

$$\mathbf{E}_{xy} = \mathbf{V} \begin{bmatrix} l_1 & & & & & & \\ & l_2 & & & \mathbf{O} & & \\ & & \ddots & & & & \\ & & & l_K & & & \\ & & & & 0 & & \\ & \mathbf{O} & & & & \ddots & \\ & & & & & & 0 \end{bmatrix} \mathbf{V}^\dagger \tag{2.164}$$

where $l_i > 0$ and $\mathbf{V}\mathbf{V}^\dagger = \mathbf{I}_M$.

In the TLS-ESPRIT formulation, the next step is to find a $2K \times K$ full rank matrix $\mathbf{W}$ such that [45]

$$\begin{bmatrix} \mathbf{E}_x & \mathbf{E}_y \end{bmatrix} \mathbf{W} = \mathbf{O}. \tag{2.165}$$

In that case from (2.162) we have

$$\mathbf{A}\begin{bmatrix} \mathbf{C} & \mathbf{B}\mathbf{C} \end{bmatrix} \mathbf{W} = \mathbf{O}. \tag{2.166}$$

Since $\mathbf{A}$ is $M \times K$ and of rank K, (2.166) is equivalent to

$$\begin{bmatrix} \mathbf{C} & \mathbf{B}\mathbf{C} \end{bmatrix} \mathbf{W} = \mathbf{O}. \tag{2.167}$$

Once again, partition $\mathbf{W}$ as

$$\mathbf{W} = \begin{bmatrix} \mathbf{W}_1 \\ \mathbf{W}_2 \end{bmatrix} \tag{2.168}$$

where $\mathbf{W}_1$ and $\mathbf{W}_2$ are two $K \times K$ matrices. This, together with (2.167) gives

$$\begin{bmatrix} \mathbf{C} & \mathbf{B}\mathbf{C} \end{bmatrix} \begin{bmatrix} \mathbf{W}_1 \\ \mathbf{W}_2 \end{bmatrix} = \mathbf{C}\mathbf{W}_1 + \mathbf{B}\mathbf{C}\mathbf{W}_2 = \mathbf{O}$$

or

$$-\mathbf{W}_1\mathbf{W}_2^{-1} = \mathbf{C}^{-1}\mathbf{B}\mathbf{C}. \tag{2.169}$$

Thus, any $\mathbf{W}$ satisfying (2.165) has the interesting property that the eigenvalues of the matrix $-\mathbf{W}_1\mathbf{W}_2^{-1}$ generated from the partition in (2.168), are given by $e^{-j\omega_k}$, $k = 1, 2, \cdots, K$. Again, the angles-of-arrival are obtained directly.

To complete this analysis, it is sufficient to exhibit such a $\mathbf{W}$ and toward this purpose, a reexamination of (2.164) shows

$$\mathbf{E}_{xy} \, \mathbf{v}_i = l_i \, \mathbf{v}_i = 0, \quad K < i \leq 2K \tag{2.170}$$

where $\mathbf{v}_i$ represents the ith column vector of $\mathbf{V}$. Since $[\mathbf{E}_x \; \mathbf{E}_y]$ is also of rank K, from (2.163), (2.170) reduces to

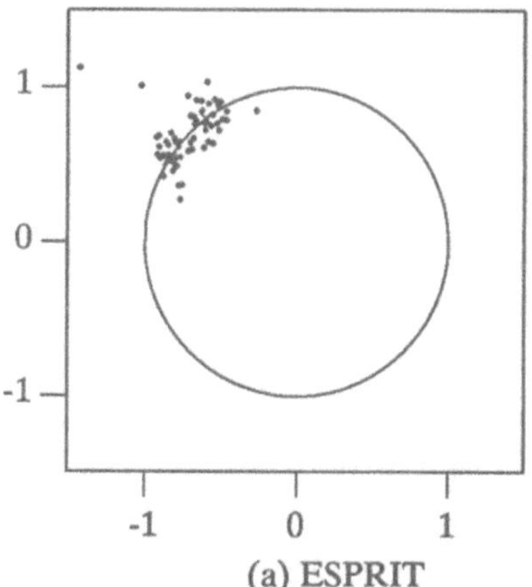
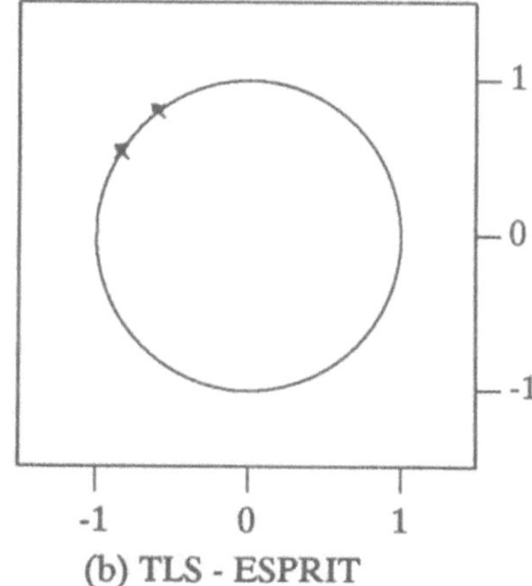

(a) ESPRIT   (b) TLS - ESPRIT

Fig. 2.15 Simulation results for two equipowered uncorrelated sources located along 35° and 45°. A seven-element array is used to receive signals. Input SNR is taken to be 15dB (number of simulations = 30, number of samples = 100).

$$\begin{bmatrix} \mathbf{E}_x & \mathbf{E}_y \end{bmatrix} \mathbf{v}_i = 0, \quad K < i \leq 2K. \tag{2.171}$$

Thus the desired **W** is given by

$$\mathbf{W} = \begin{bmatrix} \mathbf{v}_{K+1}, \mathbf{v}_{K+2}, \cdots, \mathbf{v}_{2K} \end{bmatrix} \triangleq \begin{bmatrix} \mathbf{V}_{12} \\ \mathbf{V}_{22} \end{bmatrix} \tag{2.172}$$

and the eigenvalues of $-\mathbf{V}_{12}\mathbf{V}_{22}^{-1}$ gives the actual directions-of-arrival. Fig. 2.15 shows results of simulation in an uncorrelated situation for ESPRIT and TLS-ESPRIT schemes with details as indicated there.

Though TLS-ESPRIT is superior in its performance compared to ESPRIT, it is computationally much more complex. However, computational simplicity can be maintained without sacrificing superior performance, and in that respect GEESE (GEneralized Eigenvalues utilizing Signal Subspace Eigenvectors) technique [46, 47] outperforms the former two methods. This is carried out by observing a well known property of the signal subspace; i.e.,the subspace spanned by the true direction vectors is identically the same as the one spanned by the eigenvectors corresponding to all, except the smallest set of repeating eigenvalue of the array output covariance matrix. This elementary observation forms the basis for the GEESE technique described below.

GEESE

As remarked before, a completely different point of view can be developed using (2.64). Since the K true direction vectors $\mathbf{a}(\omega_1)$, $\mathbf{a}(\omega_2)$, $\cdots$, $\mathbf{a}(\omega_K)$ are linearly independent, they span a K dimensional proper subspace called the signal subspace. Further, from (2.64), this subspace is orthogonal to the subspace spanned by the eigenvectors $\boldsymbol{\beta}_{K+1}$, $\boldsymbol{\beta}_{K+2}$, $\cdots$, $\boldsymbol{\beta}_M$, implying that the signal subspace spanned by $\mathbf{a}(\omega_1)$, $\mathbf{a}(\omega_2)$, $\cdots$, $\mathbf{a}(\omega_K)$ coincides with that spanned by the eigenvectors $\boldsymbol{\beta}_1$, $\boldsymbol{\beta}_2$, $\cdots$, $\boldsymbol{\beta}_K$. Using this crucial observation, the eigenvectors $\boldsymbol{\beta}_1$, $\boldsymbol{\beta}_2$, $\cdots$, $\boldsymbol{\beta}_K$ in the signal subspace can be expressed as a linear combination of the true direction vectors (columns of $\mathbf{A}$); i.e.,

$$\boldsymbol{\beta}_i = \sum_{k=1}^{K} \tilde{c}_{ki}\, \mathbf{a}(\omega_k), \quad i = 1, 2, \cdots, K. \tag{2.173}$$

Define the $M \times K$ signal subspace eigenvector matrix as

$$\tilde{\mathbf{B}} \triangleq \left[\boldsymbol{\beta}_1, \boldsymbol{\beta}_2, \cdots, \boldsymbol{\beta}_K \right]. \tag{2.174}$$

Using (2.173)

$$\tilde{\mathbf{B}} = \left[\sum_{k=1}^{K} \tilde{c}_{k1}\mathbf{a}(\omega_k), \sum_{k=1}^{K} \tilde{c}_{k2}\mathbf{a}(\omega_k), \cdots, \sum_{k=1}^{K} \tilde{c}_{kK}\, \mathbf{a}(\omega_k) \right] = \mathbf{A}\tilde{\mathbf{C}}$$

where $\mathbf{A}$ is as before and $\tilde{\mathbf{C}}$ is a $K \times K$ nonsingular matrix whose $(i,j)^{\text{th}}$ element is $\tilde{c}_{ij}/\sqrt{M}$. Further, define two matrices $\tilde{\mathbf{B}}_1$ and $\tilde{\mathbf{B}}_2$ using the first J rows and the 2nd to $(J+1)^{\text{th}}$ rows of $\tilde{\mathbf{B}}$ respectively where $K \leq J \leq M-1$; i.e.,

$$\tilde{\mathbf{B}}_1 = \left[\mathbf{I}_J \quad \mathbf{O}_{J,M-J} \right] \tilde{\mathbf{B}} \tag{2.175}$$

and

$$\tilde{\mathbf{B}}_2 = \left[\mathbf{O}_{J,1} \quad \mathbf{I}_J \quad \mathbf{O}_{J,M-J-1} \right] \tilde{\mathbf{B}}. \tag{2.176}$$

Then, we have the following interesting result [46].

Theorem : Let γ_i represent the generalized singular values associated with the matrix pencil $\{\,\tilde{B}_1,\,\tilde{B}_2\,\}$. Then

$$\gamma_k = e^{j\omega_k}, \quad k = 1, 2, \cdots, K. \tag{2.177}$$

Proof : From (2.175) and (2.176),

$$\tilde{B}_1 = A_1\,\tilde{C}, \quad \tilde{B}_2 = A_2\,\tilde{C} \tag{2.178}$$

where

$$A_1 = \begin{bmatrix} 1 & 1 & \cdots & 1 \\ \nu_1 & \nu_2 & \cdots & \nu_K \\ \vdots & \vdots & & \vdots \\ \nu_1^{J\text{-}1} & \nu_2^{J\text{-}1} & \cdots & \nu_K^{J\text{-}1} \end{bmatrix}$$

and

$$A_2 = \begin{bmatrix} \nu_1 & \nu_2 & \cdots & \nu_K \\ \nu_1^2 & \nu_2^2 & \cdots & \nu_K^2 \\ \vdots & \vdots & & \vdots \\ \nu_1^J & \nu_2^J & \cdots & \nu_K^J \end{bmatrix} = A_1\,B$$

with B as in (2.109). Notice that A_1, A_2 are matrices of size $J \times K$ and B is of size $K \times K$. To obtain the generalized singular values for the matrix pencil $\{\,\tilde{B}_1,\,\tilde{B}_2\,\}$, using the above representation, we have

$$\tilde{B}_1 - \gamma\tilde{B}_2 = A_1\,\tilde{C} - \gamma A_1 B\,\tilde{C} = A_1(I_K - \gamma B)\,\tilde{C} \tag{2.179}$$

Since the K columns of $\tilde{B}$ are independent, $\tilde{B}$ is of rank K $(M > K)$. Moreover, from the definitions of the rectangular matrices $\tilde{B}_1$, $\tilde{B}_2$ in (2.175), (2.176), these matrices are also of rank K (full column rank) and using (2.178), rank (A_1) = rank $(\tilde{C})$ = K since $J \geq K$. Thus, from

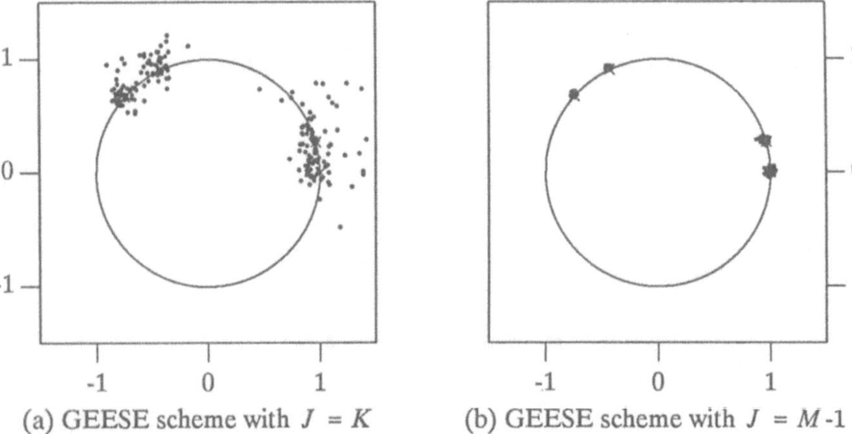

(a) GEESE scheme with $J = K$ (b) GEESE scheme with $J = M$ -1

Fig. 2.16 Simulation results for a mixed-source scene. The two sources located along 40°, 50° are uncorrelated and the two along 85°, 95° are correlated with correlation coefficient equal to $0.24 - j\,0.65$. A seven-element array is used to receive the signals. The input SNR is taken to be 12dB (number of simulations = 50, number of samples = 100).

(2.179), the singular values of the above matrix pencil $\{\tilde{\mathbf{B}}_1, \tilde{\mathbf{B}}_2\}$ are given by the roots of

$$| \mathbf{I}_K - \gamma\mathbf{B} | = 0. \tag{2.180}$$

These generalized singular values correspond to the complex conjugates of the diagonal elements of $\mathbf{B}$; i.e.,

$$\gamma_k = e^{j\omega_k}, \quad k = 1, 2, \cdots, K. \tag{2.181}$$

This completes the proof. Notice that J can be any integer between K and $M - 1$. It may be remarked that the underlying rotational invariance idea is basic to the ESPRIT scheme and in that sense all these algorithms are equivalent when covariances are exactly known.

Fig. 2.16 presents simulation results in a four source scene for the GEESE scheme with $J = K$ and $J = M - 1$. The improvement in resolution for the latter case $(J = M - 1)$ can be best explained in terms of its resolution threshold analysis discussed in chapter 3.

2.4.3 Direction Finding Using First Order Statistics

In previous sections we concentrated on methods that make use of the array output covariance matrix to determine the directions of arrival of multiple signals. By exploiting specific eigenstructure properties of this array output covariance matrix, in the case of independent and identically distributed sensor noises, these techniques have been shown to yield high resolution even when the signal sources are only partially correlated. However, when some of the signals are coherent, these covariance-based methods face serious difficulties. In addition to this, the direct use of the array output covariance matrix in direction finding has other disadvantages as well. A possibly significant one stems from the fact that the array output data, which contains information regarding all of the useful spatial parameters of interest, is essentially being squared to generate the output covariances, which are used in turn to estimate the parameters of interest. Clearly the squaring process of the output data array to generate the covariance matrix results in an unnecessarily large dynamic range for the processor, and hence it will be highly desirable to rephrase the direction finding problem in terms of the output data matrix itself.

We explore this possibility here and show that it is indeed possible to estimate all arrival angles of the targets from suitable processing on the array output data matrix itself [48]. In addition to a lower dynamic range for the processor, this technique is shown to be insensitive to the degree of correlation of the signal sources, thereby making it an attractive alternative in coherent situations. Moreover, in this case there are no restrictions on the structure of the noise covariance matrix and consequently the sensor noises can be dependent with unequal variances.

Consider the scenario described in section 2.3. No assumptions are made regarding the coherence among these signals and in particular they all may be coherent as happens in a multiple echo situation of a single source. It is assumed that the signals and noises are uncorrelated stationary processes, and further the noises are assumed to be zero mean with an arbitrary noise covariance matrix. From (2.49), we obtain

$$a_i \triangleq E[x_i(t)] = \sum_{k=1}^{K} \mu_k \, e^{-j\pi d_i \cos\theta_k} , \quad i = 0, 1, \cdots, (M\text{-}1) , \quad (2.182)$$

where

$$\mu_k \triangleq E[u_k(t)], \quad k = 1, 2, \cdots, K. \tag{2.183}$$

We will assume that these mean values are nonzero. When this assumption holds true, the situation here is identical to that in (2.84) in section 2.3.2 with $\sigma_0^2 = 0$. Proceeding as before, *if all μ_k's are real*, then in the case of a uniform array, the hermitian Toeplitz matrix

$$\mathbf{T}_1 \triangleq \begin{bmatrix} a_0 & a_1^* & a_2^* & \cdots & a_{M-1}^* \\ a_1 & a_0 & a_1^* & \cdots & a_{M-2}^* \\ a_2 & a_1 & a_0 & \cdots & a_{M-3}^* \\ \vdots & \vdots & \vdots & \ddots & \vdots \\ a_{M-1} & a_{M-2} & a_{M-3} & \cdots & a_0 \end{bmatrix} \tag{2.184}$$

has the same structural properties as in (2.86). Thus, we can write

$$\mathbf{T}_1 = \mathbf{A}\mathbf{R}_\mu\mathbf{A}^\dagger \tag{2.185}$$

where $\mathbf{A}$ is as in (2.55) and

$$\mathbf{R}_\mu = diag[\,\mu_1, \mu_2, \cdots, \mu_K\,]. \tag{2.186}$$

Since the direction vectors are independent, $\mathbf{A}$, and therefore $\mathbf{T}_1$, are both of rank K. This in turn implies that $(M\text{-}K)$ eigenvalues of $\mathbf{T}_1$ are zeros and further the corresponding eigenvectors $\mathbf{u}_{(1)}, \mathbf{u}_{(2)}, \cdots,$ $\mathbf{u}_{(M\text{-}K)}$ are orthogonal to the direction vectors $\mathbf{a}(\omega_1), \mathbf{a}(\omega_2), \cdots,$ $\mathbf{a}(\omega_K)$; i.e.,

$$\mathbf{u}_{(i)}^\dagger\mathbf{a}(\omega_k) = 0, \quad i = 1, 2, \cdots, M\text{-}K, \quad k = 1, 2, \cdots, K. \tag{2.187}$$

Once again the zeros of the function

$$D(\omega) = \sum_{i=1}^{M\text{-}K} |\,\mathbf{u}_{(i)}^\dagger\mathbf{a}(\omega)\,|^2 \tag{2.188}$$

will correspond to the true directions of arrival. The μ_ks can be easily transformed into equivalent real quantities (see section 2.3.2) by expanding the M element array by adding a new set of $(M\text{-}1)$ elements symmetrically about the reference point (see Fig. 2.11). Then the output at the ith element of this new set of elements will be

$$x_{-i}(t) = \sum_{k=1}^{K} u_k(t) e^{j\pi d_i \cos\theta_k} + n_{-i}(t), \quad i = 1, 2, \cdots, (M\text{-}1) \quad (2.189)$$

and by defining

$$a_{-i} \triangleq E[x_{-i}(t)]$$

we have

$$a_{-i} = \sum_{k=1}^{K} \mu_k e^{j\pi d_i \cos\theta_k}, \quad i = 1, 2, \cdots, (M\text{-}1). \quad (2.190)$$

Now let

$$c_i = \frac{a_i^* + a_{-i}}{2}. \quad (2.191)$$

Using (2.182) and (2.190) in (2.191), we have

$$c_i = \sum_{k=1}^{K} b_k e^{j\pi d_i \cos\theta_k}, \quad i = 1, 2, \cdots, (M\text{-}1) \quad (2.192)$$

where by definition

$$b_k = \text{Re}(\mu_k) \neq 0, \quad k = 1, 2, \cdots, K \quad (2.193)$$

and this together with a_0 given by (2.182) allows us to define

$$c_0 = \text{Re}(a_0) = \sum_{k=1}^{K} b_k \neq 0. \quad (2.194)$$

Then the hermitian Toeplitz matrix $\mathbf{T}_2$ formed from these mean values has the form

$$
\mathbf{T}_2 = \begin{bmatrix}
c_0 & c_1 & c_2 & \cdots & c_{M-1} \\
c_1^* & c_0 & c_1 & \cdots & c_{M-2} \\
c_2^* & c_1^* & c_0 & \cdots & c_{M-3} \\
\vdots & \vdots & \vdots & \ddots & \vdots \\
c_{M-1}^* & c_{M-2}^* & c_{M-3}^* & \cdots & c_0
\end{bmatrix} \qquad (2.195)
$$

and can be written as

$$\mathbf{T}_2 = \mathbf{A}\,\mathbf{B}_0\,\mathbf{A}^\dagger$$

with

$$\mathbf{B}_0 = diag\,[b_1, b_2, \cdots, b_K]\,.$$

Note that b_k, $k = 1, 2, \cdots, K$ being real, $\mathbf{T}_2$ is of rank K, and hence allows the following representation

$$\mathbf{T}_2 = \mathbf{U}\mathbf{\Lambda}\mathbf{U}^\dagger$$

where

$$\mathbf{U} = \left[\mathbf{u}_1, \mathbf{u}_2, \cdots, \mathbf{u}_K, \mathbf{u}_{(1)}, \mathbf{u}_{(2)}, \cdots, \mathbf{u}_{(M-K)}\right] \qquad (2.196)$$

$$\mathbf{\Lambda} = diag\,[\lambda_1, \lambda_2, \cdots, \lambda_K, 0, 0, \cdots, 0]$$

with

$$\mathbf{U}\mathbf{U}^\dagger = \mathbf{I}$$

and satisfying $u_{ii} \geq 0$, $i = 1, 2, \cdots, M$ for uniqueness. Here $\mathbf{u}_1$, $\mathbf{u}_2$, $\cdots$, $\mathbf{u}_K$, are eigenvectors corresponding to the distinct nonzero eigenvalues λ_1, λ_2, $\cdots$, λ_K, and the set $\mathbf{u}_{(i)}$, $i = 1, 2, \cdots, M\text{-}K$, of eigenvectors correspond to the repeating zero eigenvalue. From the above discussion it is also clear that the $(M\text{-}K)$ eigenvectors of $\mathbf{T}_2$ that correspond to its repeating zero eigenvalue are orthogonal to the actual direction vectors $\mathbf{a}(\omega_k)$, $k = 1, 2, \cdots, K$, and hence (2.188) can be applied to estimate the signal directions of arrival. It may be remarked that when the signal mean values have positive real parts (i.e.; b_k, $k = 1, 2, \cdots, K$ in (2.193) are all positive), $\mathbf{T}_2$ is a

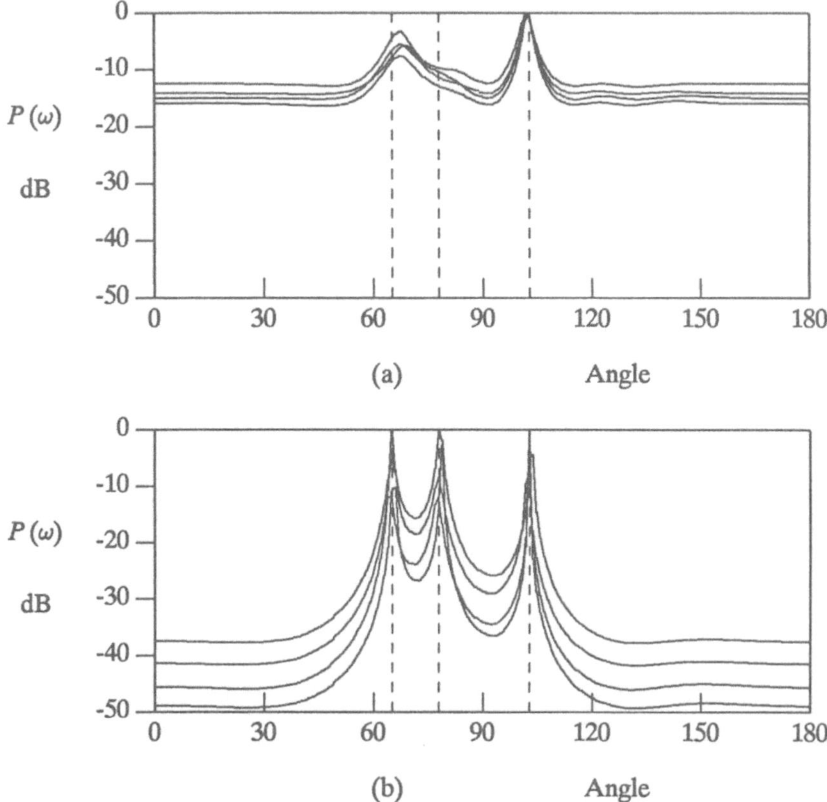

Fig. 2.17 Direction finding from first order statistics. (a) $P(\omega)$ using the conventional MUSIC algorithm. (b) $P(\omega)$ using the proposed mean based scheme $(P(\omega) = 1/D(\omega))$.

nonnegative definite matrix and the above $(M\text{-}K)$ eigenvectors will correspond to the lowest eigenvalue, which is zero in this case. In all these events, the K zeros of $D(\omega)$ given by (2.188) will correspond to the true arrival angles.

Simulation results are presented in Figs. 2.17 and 2.18 to illustrate the performance of the array output mean vector based scheme and to compare it with the conventional eigenstructure-based techniques.

Fig. 2.17 represents a correlated source scene in presence of unequal sensor noises. A nine-element uniform array receives signals

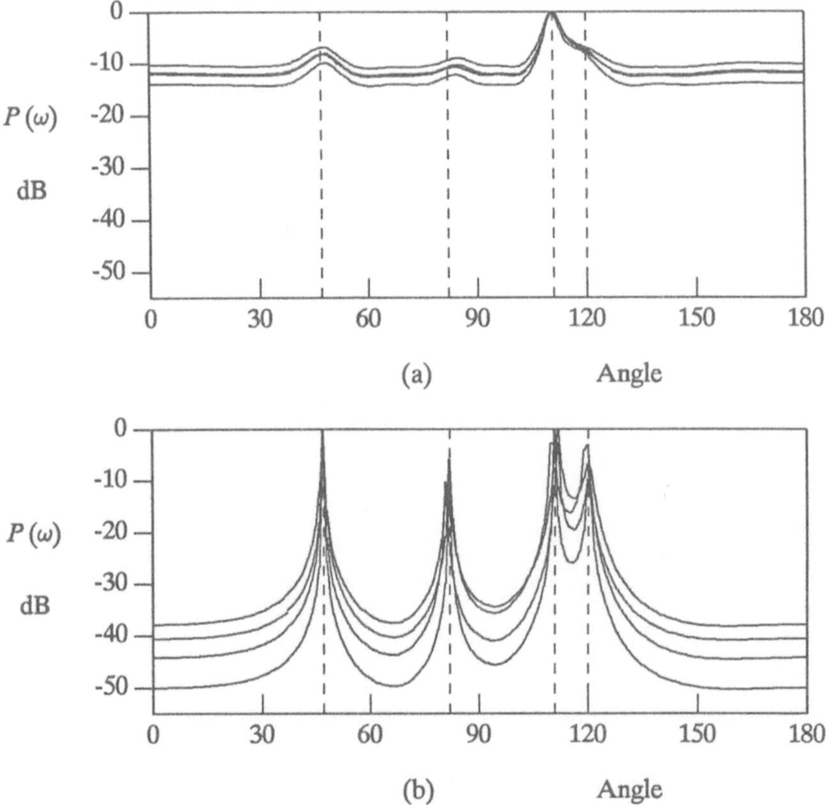

Fig. 2.18 A correlated/coherent source scene with correlated sensor noises. An eleven-element uniform array receives signals from four correlated sources, of which two are completely coherent. The arrival angles are 47°, 82°, 111°, and 120°. The input SNR of the reference signal (along $\theta_4 = 120°$) is 7 dB with respect to the third sensor element. The normalized noise variances are 3.0, 0.3, 1.0, 0.5, 2.0, 0.25, 4.0, 0.125, 8.0, 0.5, and 2.0 and their interelement noise correlations are as given in Table 2.3. One hundred fifty samples are used to estimate the array output mean vector. Each graph is separately normalized by the maximum value in that part. (a) $P(\omega)$ using the conventional MUSIC algorithm. (b) $P(\omega)$ using the proposed mean based scheme.

Table 2.3 Interelement sensor noise correlation coefficients (see Fig. 2.18).

	1	2	3	4	5
2	-j0.71				
3	-0.5 -j0.5	0.35-j0.71			
4	-0.27-j0.54	0.46-j0.50	0.76+j0.32		
5	-0.21-j0.43	0.36-j0.39	0.60+j0.26	0.79	
6	0.08-j0.32	0.37+j0.20	0.22+j0.30	-0.02-j0.08	-0.02-j0.31
7	-0.18-j0.37	0.31-j0.34	0.51+j0.22	0.68	0.32-j0.22
8	-j0.34	0.49	0.09+j0.17	0.12+j0.05	0.20+j0.14
9	-0.16-j0.33	0.28-j0.30	0.46+j0.2	0.61	0.28-j0.2
10	0.06-j0.25	0.29+j0.16	0.17+j0.24	-0.02-j0.07	-0.01-j0.24
11	0.15+j0.15	0.11+j0.11	-0.30	-0.25+j0.08	-0.02+j0.01

	6	7	8	9	10
7	0.13-j0.24				
8	0.26-j0.03	-0.17+j0.03			
9	0.12-j0.21	0.89	-0.21-j0.14		
10	0.78	0.10+j0.07	0.42+j0.13	j0.08	
11	0.02	-0.06+j0.12	0.19-j0.18	0.09+j0.04	0.08+j0.10

from three correlated sources with power levels 1.0, 0.5, 2.0 and correlation coefficients $\rho_{12} = 0.42+j0.57$, $\rho_{13} = -0.35-j0.30$, $\rho_{23} = -0.53+j0.49$, respectively. The signal arrival angles are 65°, 78°, and 103° and their respective mean values are $0.50+j0.50$, $0.35-j0.35$, and $0.96+j0.29$. The input SNR of the reference signal (along $\theta_1 = 65°$) is 4 dB with respect to the first sensor element. The normalized noise variances are 1.0, 0.5, 2.0, 0.25, 4.0, 0.125, 8.0, 0.5, and 2.0. One hundred samples are used to estimate the array output mean vector and covariance matrix. The application of the MUSIC method [19] to the estimated covariance matrix resulted in Fig. 2.17.(a) and the array output mean vector based scheme resulted in Fig. 2.17.(b). All three directions of arrival can be unambiguously identified in the latter case.

Fig. 2.18 represents a mixed source scene consisting of three partially correlated sources of which one undergoes multipath reflection, generating an additional coherent signal. The signal power levels are 1.5, 0.5, 2.0, 1.0 with correlation coefficients ρ_{12} = $0.60+j\,0.80$, ρ_{13} = $-0.47-j\,0.39$, ρ_{14} = $0.81+j\,0.12$, ρ_{23} = $-0.60+j\,0.14$, ρ_{24} = $0.58-j\,0.57$ and ρ_{34} = $-0.59-j\,0.03$. The signal mean values are given by $0.52-j\,0.42$, $0.29+j\,0.26$, $0.73+j\,0.24$, and $0.49-j\,0.24$. The input SNR of the reference signal (along θ_4 = 120°) is 7 dB with respect to the third sensor element. The noises have unequal variances and are correlated with each other. Their interelement noise correlations are as given in Table 2.3. The improvement in performance of the proposed scheme in terms of signal resolvability, irrespective of the signal coherence and interelement noise correlation, is visible in all these cases. It may be remarked that the nonzero assumption in (2.193), which treats the signals to have nonzero mean is crucial for this method.

Appendix 2.A
Coherent and Correlated Signal Scene

We will demonstrate here that the forward/backward smoothing scheme discussed in section 2.3.3 readily extends to the general situation where the source scene consists of $K+J$ signals $u_1(t)$, $u_2(t)$, $\cdots$, $u_K(t)$, $u_{K+1}(t)$, $\cdots$, $u_{K+J}(t)$, of which the first K signals are completely coherent and the last $(J+1)$ signals are partially correlated. Thus the coherent signals are partially correlated with the remaining set of signals. Further, the respective arrival angles are assumed to be θ_1, θ_2, $\cdots$, θ_K, θ_{K+1}, $\cdots$, θ_{K+J}. As before, the signals are taken to be uncorrelated with the noise and the noise is assumed to be identical and uncorrelated from element to element. Using (2.68) the output $x_i(t)$ of the ith sensor element at time t in this case can be written as

$$x_i(t) = u_1(t) \sum_{k=1}^{K} \alpha_k e^{-j(i-1)\omega_k} + \sum_{k=K+1}^{K+J} u_k(t) e^{-j(i-1)\omega_k} + n_i(t),$$

$$i = 1, 2, \cdots, M . \qquad (2.A.1)$$

With x(t) as in (2.51), this gives

$$\mathbf{x}(t) = \tilde{\mathbf{A}}\,\mathbf{v}(t) + \mathbf{n}(t)\,, \tag{2.A.2}$$

where $\tilde{\mathbf{A}}$ is as given by (2.88) and

$$\mathbf{v}(t) = \begin{bmatrix} \mathbf{u}_1(t) \\ \mathbf{u}_2(t) \end{bmatrix}. \tag{2.A.3}$$

Here

$$\mathbf{u}_1(t) = \Big[\, u_1(t), u_2(t), \cdots, u_K(t) \,\Big]^{T} = u_1(t)\boldsymbol{\alpha} \tag{2.A.4}$$

with $\boldsymbol{\alpha}$ as in (2.103) and

$$\mathbf{u}_2(t) = \Big[\, u_{K+1}(t), u_{K+2}(t), \cdots, u_{K+J}(t) \,\Big]^{T}. \tag{2.A.5}$$

Following (2.107) - (2.113), (2.116) - (2.121) and (2.127), the forward/backward smoothed covariance matrix $\tilde{\mathbf{R}}$ in this case can be written as

$$\tilde{\mathbf{R}} = \tilde{\mathbf{A}}\,\tilde{\mathbf{R}}_v\,\tilde{\mathbf{A}}^{\dagger} + \sigma^2\mathbf{I}\,, \tag{2.A.6}$$

where

$$\tilde{\mathbf{R}}_v = \frac{1}{2L}\sum_{l=1}^{L}\tilde{\mathbf{B}}^{l-1}(\mathbf{R}_v + \mathbf{R}_{\tilde{v}})(\tilde{\mathbf{B}}^{l-1})^{\dagger}. \tag{2.A.7}$$

It remains to show that $\tilde{\mathbf{R}}_v$ is of full rank irrespective of the coherency among some of the arrivals. Here

$$\tilde{\mathbf{B}} = \begin{bmatrix} \mathbf{B}_1 & \mathbf{O} \\ \mathbf{O} & \mathbf{B}_2 \end{bmatrix}, \tag{2.A.8}$$

where $\mathbf{B}_1 = \mathbf{B}$, with $\mathbf{B}$ as in (2.109) and

$$\mathbf{B}_2 = diag\,[\nu_{K+1}, \nu_{K+2}, \cdots, \nu_{K+J}] \tag{2.A.9}$$

with $\nu_k\,;\, k = 1, 2, \cdots, K+J$ as given by (2.110) and

$$\mathbf{R}_v \triangleq E[\mathbf{v}(t)\mathbf{v}^\dagger(t)] = \begin{bmatrix} \mathbf{R}_{11} & \mathbf{R}_{12} \\ \\ \mathbf{R}_{12}^\dagger & \mathbf{R}_{22} \end{bmatrix}. \tag{2.A.10}$$

Using (2.A.3) - (2.A.5) and $E[\,|u_1(t)|^2] = 1$, it is easy to see that

$$\mathbf{R}_{11} = E[\mathbf{u}_1(t)\mathbf{u}_1^\dagger(t)] = \boldsymbol{\alpha}\boldsymbol{\alpha}^\dagger \tag{2.A.11}$$

Similarly

$$\mathbf{R}_{12} = E[\mathbf{u}_1(t)\mathbf{u}_2^\dagger] = \boldsymbol{\alpha}\boldsymbol{\gamma}^\dagger \tag{2.A.12}$$

with

$$\boldsymbol{\gamma} = \left[\gamma_1, \gamma_2, \cdots, \gamma_J\right]^T, \tag{2.A.13}$$

where

$$\gamma_i = E[u_1(t)u_{K+i}^*(t)], \quad i = 1, 2, \cdots, J, \tag{2.A.14}$$

and

$$\mathbf{R}_{22} = E[\mathbf{u}_2(t)\mathbf{u}_2^\dagger(t)]. \tag{2.A.15}$$

From the partially correlated assumption among the later J signals, it follows that their correlation matrix $\mathbf{R}_{22}$ is of full rank and hence it has the representation

$$\mathbf{R}_{22} = \mathbf{A}\mathbf{A}^\dagger \tag{2.A.16}$$

where $\mathbf{A}$ is again a full rank matrix of size $J \times J$. In a similar manner following (2.119), $\mathbf{R}_{\tilde{v}}$ can be written as

$$\mathbf{R}_{\tilde{v}} = \tilde{\mathbf{B}}^{-(M_0-1)}\mathbf{R}_v^*\left(\tilde{\mathbf{B}}^{-(M_0-1)}\right)^\dagger = \begin{bmatrix} \tilde{\mathbf{R}}_{11} & \tilde{\mathbf{R}}_{12} \\ \\ \tilde{\mathbf{R}}_{12}^\dagger & \tilde{\mathbf{R}}_{22} \end{bmatrix}, \tag{2.A.17}$$

and proceeding as before,

$$\tilde{\mathbf{R}}_{11} = \delta\,\delta^{\dagger} \tag{2.A.18}$$

with δ as in (2.123) and

$$\tilde{\mathbf{R}}_{12} = \delta\,\tilde{\gamma}^{\dagger} \tag{2.A.19}$$

with

$$\tilde{\gamma} = \left[\tilde{\gamma}_1, \tilde{\gamma}_2, \cdots, \tilde{\gamma}_J\right]^{T}, \tag{2.A.20}$$

where

$$\tilde{\gamma}_i = \gamma_i^{*}\,(\nu_{K+i})^{M_0-1}, \quad i = 1, 2, \cdots, J. \tag{2.A.21}$$

Here γ_i is as defined in (2.A.14) and ν_{K+i} is obtained by extending the definition in (2.110). Further,

$$\tilde{\mathbf{R}}_{22} = \mathbf{B}_2^{-(M_0-1)}\,\mathbf{R}_{22}\left(\mathbf{B}_2^{-(M_0-1)}\right)^{\dagger} = \tilde{\mathbf{A}}\,\tilde{\mathbf{A}}^{\dagger} \tag{2.A.22}$$

with

$$\tilde{\mathbf{A}} = \mathbf{B}_2^{-(M_0-1)}\,\mathbf{A}, \tag{2.A.23}$$

where $\tilde{\mathbf{A}}$ again is a full rank matrix of size $J \times J$. With (2.A.8) - (2.A.23) in (2.A.7), it simplifies to

$$\tilde{\mathbf{R}}_v = \frac{1}{2L}\begin{bmatrix} \displaystyle\sum_{l=1}^{L} \mathbf{B}_1^{l-1}(\mathbf{R}_{11}+\tilde{\mathbf{R}}_{11})(\mathbf{B}_1^{l-1})^{\dagger} & \displaystyle\sum_{l=1}^{L} \mathbf{B}_1^{l-1}(\mathbf{R}_{12}+\tilde{\mathbf{R}}_{12})(\mathbf{B}_2^{l-1})^{\dagger} \\[2em] \displaystyle\sum_{l=1}^{L} \mathbf{B}_2^{l-1}(\mathbf{R}_{12}^{\dagger}+\tilde{\mathbf{R}}_{12}^{\dagger})(\mathbf{B}_1^{l-1})^{\dagger} & \displaystyle\sum_{l=1}^{L} \mathbf{B}_2^{l-1}(\mathbf{R}_{22}+\tilde{\mathbf{R}}_{22})(\mathbf{B}_2^{l-1})^{\dagger} \end{bmatrix}$$

$$= \frac{1}{2L}\begin{bmatrix} \mathbf{G}_1\mathbf{G}_1^{\dagger} & \mathbf{G}_1\mathbf{G}_2^{\dagger} \\[1em] \mathbf{G}_2^{\dagger}\mathbf{G}_1 & \mathbf{G}_3\mathbf{G}_3^{\dagger} \end{bmatrix} = \frac{1}{2L}\begin{bmatrix} \mathbf{G}_1 & \mathbf{O} \\[1em] \mathbf{G}_2 & \mathbf{G}_4 \end{bmatrix}\begin{bmatrix} \mathbf{G}_1^{\dagger} & \mathbf{G}_2^{\dagger} \\[1em] \mathbf{O} & \mathbf{G}_4^{\dagger} \end{bmatrix}, \tag{2.A.24}$$

where

$$G_1 = \left[\boldsymbol{\alpha}, \mathbf{B}_1 \boldsymbol{\alpha}, \cdots, \mathbf{B}_1^{L-1} \boldsymbol{\alpha}, \boldsymbol{\delta}, \mathbf{B}_1 \boldsymbol{\delta}, \cdots, \mathbf{B}_1^{L-1} \boldsymbol{\delta} \right], \quad (2.\text{A}.25)$$

$$G_2 = \left[\boldsymbol{\gamma}, \mathbf{B}_2 \boldsymbol{\gamma}, \cdots, \mathbf{B}_2^{L-1} \boldsymbol{\gamma}, \tilde{\boldsymbol{\gamma}}, \mathbf{B}_2 \tilde{\boldsymbol{\gamma}}, \cdots, \mathbf{B}_2^{L-1} \tilde{\boldsymbol{\gamma}} \right], \quad (2.\text{A}.26)$$

$$G_3 = \left[\mathbf{\Lambda}, \mathbf{B}_2 \mathbf{\Lambda}, \cdots, \mathbf{B}_2^{L-1} \mathbf{\Lambda}, \tilde{\mathbf{\Lambda}}, \mathbf{B}_2 \tilde{\mathbf{\Lambda}}, \cdots, \mathbf{B}_2^{L-1} \tilde{\mathbf{\Lambda}} \right], \quad (2.\text{A}.27)$$

and G_4 satisfies

$$G_3 G_3^\dagger = G_2 G_2^\dagger + G_4 G_4^\dagger. \quad (2.\text{A}.28)$$

Define

$$\tilde{G} = \begin{bmatrix} G_1 & \mathbf{O} \\ G_2 & G_4 \end{bmatrix}. \quad (2.\text{A}.29)$$

Then

$$\tilde{\mathbf{R}}_v = \frac{1}{2L} \tilde{G} \tilde{G}^\dagger. \quad (2.\text{A}.30)$$

Clearly the rank of $\tilde{\mathbf{R}}_v$ is the same as that of $\tilde{G}$. An examination of (2.A.25) shows that $G_1 G_1^\dagger$ is the average of the source covariance matrix corresponding to the completely coherent situation (see (2.130)) and hence from the result derived in section 2.3.3, it follows that $G_1 G_1^\dagger$ is of full rank K so long as $L \geq [K/2]$. Now it remains to show that G_4 is also of full row rank J, which together with (2.A.29) implies that $\tilde{G}$ and hence $\tilde{\mathbf{R}}_v$ is of full rank $K+J$. From (2.A.26) - (2.A.28) we have

$$G_4 G_4^\dagger = G_3 G_3^\dagger - G_2 G_2^\dagger$$

$$= \sum_{l=1}^{L} \mathbf{B}_2^{l-1} (\mathbf{\Lambda} \mathbf{\Lambda}^\dagger - \boldsymbol{\gamma} \boldsymbol{\gamma}^\dagger)(\mathbf{B}_2^{l-1})^\dagger$$

$$+ \sum_{l=1}^{L} \mathbf{B}_2^{l-1} (\tilde{\mathbf{\Lambda}} \tilde{\mathbf{\Lambda}}^\dagger - \tilde{\boldsymbol{\gamma}} \tilde{\boldsymbol{\gamma}}^\dagger)(\mathbf{B}_2^{l-1})^\dagger. \quad (2.\text{A}.31)$$

In the first summation here, $\mathbf{\Lambda}$ and γ are of ranks J and 1 respectively and hence $(\mathbf{\Lambda}\mathbf{\Lambda}^{\dagger} - \tilde{\gamma}\tilde{\gamma}^{\dagger})$ is at least of rank $J - 1$. Once again resorting to the argument used in establishing (2.135) in section 2.3.3, it follows that each summation and hence $\mathbf{G}_4$ is of full row rank J so long as $L > 1$. This establishes the nonsingularity of $\tilde{\mathbf{R}}_v$ whenever $L \geq [K/2]$. As a result, the smoothed covariance matrix $\tilde{\mathbf{R}}$ in (2.A.6) has structurally the same form as the covariance matrix for some noncoherent set of $K + J$ signals. Hence, the eigenstructure-based techniques can be applied to this smoothed matrix irrespective of the coherence of the original set of signals to successfully estimate their directions of arrival. This completes the proof.

Appendix 2.B
Program Listings

The first part of the following program generates the exact covariance matrix and their estimated counterparts for a multiple source scene with arbitrary correlation among the sources. The second part describes the various angle of arrival estimation algorithms discussed in this chapter.

```
      subroutine corex(mm,kk,m,k,sn0,a,p,d,nv,cb,cor,cr)
cccccccccccccccccccccccccccccccccccccccccccccccccccccccccccccccc
c This subroutine generates exact covariance matrix.
c
c inputs
c     kk     dimension of cb(.) and cor(.) in main program
c     mm     dimension of cr(.) in main program
c     k      number of sources
c     m      number of elements
c     sn0    reference signal power / average noise power
c     a(k)   angles of arrival in radian
c     p(k)   powers of sources / reference signal power
c     d(m)   sensor locations normalized by half wavelength
c     nv(m)      sensor noise powers / average noise power
c     cb(k,k)    relation among sources
c
c outputs
```

```fortran
c       cr(m,m)      exact covariance matrix
c       cor(k,k) correlation coefficients among sources
ccccccccccccccccccccccccccccccccccccccccccccccccccccccccccccccccccccc
c
c global variables
        double precision sn0,a(kk),p(kk),d(mm),nv(mm)
        double complex cb(kk,kk),cor(kk,kk),cr(mm,mm)
c
c local variables
        double precision pi,tmp
        double complex cw(20,20),ctmp
        pi=atan(1.d0)*4.
c
c calculate signal component of cr(.)
        call trans(kk,k,cb,cor)
        do 10 i=1,k
        do 10 j=1,k
            cw(i,j)=sqrt(p(i)*p(j))*cor(i,j)
 10     continue
        do 20 i=1,m
        do 20 j=1,i
            cr(i,j)=cmplx(0.,0.)
            do 30 ik=1,k
            do 30 lk=1,k
                tmp=d(i)*cos(a(ik))-d(j)*cos(a(lk))
                ctmp=cw(ik,lk)*cexp(cmplx(0.,pi*tmp))
                cr(i,j)=cr(i,j)+ctmp
 30         continue
            if (i.ne.j) cr(j,i)=conjg(cr(i,j))
 20     continue
c
c add noise component to cr(.)
        do 40 i=1,m
            cr(i,i)=cr(i,i)+cmplx(sn0*nv(i),0.)
 40     continue
        return
        end
c
c
```

```fortran
      subroutine corsrc(dseed,mm,kk,m,k,n,snr0,a,p,d,nv,cb,cor,cr)
ccccccccccccccccccccccccccccccccccccccccccccccccccccccccccccccccc
c This subroutine generates estimated covariance matrix
c using n samples.
c
c inputs
c     kk     dimension of cb(.) in main program
c     mm     dimension of cr(.) in main program
c     k      number of sources
c     m      number of elements
c     a(k)   angles of arrival
c     p(k)   power levels
c     d(m)   sensor locations normalized by half wavelength
c     nv(m)       noise variances of array elements
c     cb(k,k)     relations among sources
c     dseed  seed of random number generation
c
c outputs
c     cr(m,m)     output covariance matrix
c     cor(k,k) correlation coefficients among sources
ccccccccccccccccccccccccccccccccccccccccccccccccccccccccccccccccc
c
c global variables
      double precision dseed,snr0,a(kk),p(kk),d(mm),nv(mm)
      double complex cr(mm,mm),cb(kk,kk),cor(kk,kk)
c
c local variables
      double complex cx(20)
c
c clear cr(.)
      do 10 im=1,m
      do 10 jm=1,m
          cr(im,jm)=cmplx(0.,0.)
 10    continue
c
c calculate cr(.)
      call trans(kk,k,cb,cor)
      do 20 in=1,n
          call corx(dseed,mm,kk,m,k,snr0,a,p,d,cb,nv,cx)
```

```fortran
              do 30 im = 1,m
              do 30 j = 1,im
                      cr(im,j) = cr(im,j) + cx(im)*conjg(cx(j))
30            continue
20      continue
        do 40 im = 1,m
        do 40 j = 1,im
              cr(im,j) = cr(im,j)/float(n)
              if (im.ne.j) then
                      cr(j,im) = conjg(cr(im,j))
              endif
40      continue
        return
        end
c
c
        subroutine corx(dseed,mm,kk,m,k,snr0,a,p,d,cb,nv,cx)
ccccccccccccccccccccccccccccccccccccccccccccccccccccccccccccccccc
c This subroutine generates array output data.
ccccccccccccccccccccccccccccccccccccccccccccccccccccccccccccccccc
c
c global variables
        double precision dseed,snr0,d(mm),a(kk),p(kk),nv(mm)
        double complex cx(mm),cb(kk,kk)
c
c local variables
        real rsrc(40),gauss(40)
        double precision pi,tmp
        double complex cs(20),ct(20),ctmp
        pi = atan(1.d0)*4.
c
c generate gaussian random variables for signals and noises
        call ggnml(dseed,2*k,rsrc)
        call ggnml(dseed,2*m,gauss)
        do 10 ik = 1,k
              ct(ik) = cmplx(rsrc(ik),rsrc(ik+k))/sqrt(2.d0)
10      continue
c
c implement source correlations
```

```fortran
      do 20 i=1,k
          cs(i)=cmplx(0.,0.)
          do 30 j=1,i
              cs(i)=cs(i)+ct(j)*cb(i,j)
30        continue
          cs(i)=sqrt(p(i))*cs(i)
20    continue
c
c generate sensor outputs
      do 40 im=1,m
          cx(im)=cmplx(0.,0.)
          do 50 ik=1,k
              tmp=cmplx(0.,pi*d(im)*cos(a(ik)))
              ctmp=cexp(cmplx(0.,tmp))*cs(ik)
              cx(im)=cx(im)+ctmp
50        continue
c
c add noise components
          ctmp=cmplx(gauss(im),gauss(m+im))
          cx(im)=cx(im)+snr0*sqrt(nv(im))*ctmp
40    continue
      return
      end
c
c
      subroutine trans(kk,k,cb,cor)
ccccccccccccccccccccccccccccccccccccccccccccccccccccccccccccccccccccc
c This subroutine calculates correlation coefficients.
ccccccccccccccccccccccccccccccccccccccccccccccccccccccccccccccccccccc
c
c global variable
      double complex cb(kk,kk),cor(kk,kk)
c
c local variable
      double precision tmp
      double complex ctmp
c
c normalize cb(.)
      do 10 i=1,k
```

```fortran
              tmp=0.
              do 20 j=1,i
                      tmp=tmp+abs(cb(i,j))**2
20            continue
              tmp=sqrt(tmp)
              do 30 j=1,i
                      cb(i,j)=cb(i,j)/tmp
30            continue
              cb(i,i)=cmplx(abs(cb(i,i)),0.)
10    continue
c
c calculate correlation coefficients
      do 40 j=1,k
      do 40 i=1,j
              if (i.eq.j) then
                      cor(i,j)=cmplx(1.,0.)
              else
                      ctmp=cmplx(0.,0.)
                      do 50 m=1,i
                              ctmp=ctmp+cb(i,m)*conjg(cb(j,m))
50                    continue
                      cor(i,j)=ctmp
                      cor(j,i)=conjg(ctmp)
              endif
40    continue
      return
      end
c
c
      subroutine beamform(mm,m,d,cr)
ccccccccccccccccccccccccccccccccccccccccccccccccccccccccccccccccccc
c This subroutine calculates beamformer power estimator.
c
c inputs
c      mm   dimension of cr(.) in main program
c      m    number of elements
c      d(m) sensor locations normalized by half wavelength
c      cr(m,m)    array output covariance matrix
c
```

```
c output      P(w) as a function of w
cccccccccccccccccccccccccccccccccccccccccccccccccccccccccccccccccc
c
c global variables
      double precision d(mm)
      double complex cr(mm,mm)
c
c local variables
      double precision pi,t
      double complex ctmp
      pi=atan(1.d0)*4.
c
c calculate P(w)
      do 10 i=0,180
            t=float(i)*pi/180.
            call strxxs(mm,m,t,d,cr,ctmp)
            print*,i,real(ctmp)
  10    continue
      return
      end
c
c
      subroutine capon(mm,m,d,cr)
cccccccccccccccccccccccccccccccccccccccccccccccccccccccccccccccccc
c This subroutine calculates Capon's minimum variance power
c estimator.
c
c inputs
c      mm    dimension of cr(.) in main program
c      m     number of elements
c      d(m)  sensor locations normalized by half wavelength
c      cr(m,m)    array output covariance matrix
c
c output      P(w) as a function of w
cccccccccccccccccccccccccccccccccccccccccccccccccccccccccccccccccc
c
c global variables
      double precision d(mm)
      double complex cr(mm,mm)
```

```fortran
c
c local variables
      double precision pi,t,wk(20)
      double complex ctmp,cinv(20,20)
      pi=atan(1.d0)*4.
c
c inverse Rxx
      call inverse(mm,m,cr,cinv,wk)
c
c calculate P(w)
      do 10 i=0,180
            t=float(i)*pi/180.
            call strxxs(mm,m,t,d,cinv,ctmp)
            ctmp=1./ctmp
            print*,i,real(ctmp)
  10    continue
      return
      end
c
c
      subroutine lp(mm,m,d,cr)
ccccccccccccccccccccccccccccccccccccccccccccccccccccccccccccccccccccc
c This subroutine calculates linear prediction power estimator.
c
c inputs
c      mm   dimension of cr(.) in main program
c      m      number of elements
c      d(m)  sensor locations normalized by half wavelength
c      cr(m,m)     array output covariance matrix
c
c output     P(w) as a function of w
ccccccccccccccccccccccccccccccccccccccccccccccccccccccccccccccccccccc
c
c global variables
      double precision d(mm)
      double complex cr(mm,mm)
c
c local variables
      double precision pi,t,wk(20),tmp
```

```
      double complex ctmp,cinv(20,20),ca(20)
      pi=atan(1.d0)*4.
c
c inverse Rxx
      call inverse(mm,m,cr,cinv,wk)
      do 10 i=1,m
           ca(i)=cinv(i,1)
 10     continue
c
c calculate P(w)
      do 20 i=0,180
           t=float(i)*pi/180.
           call atmuls(mm,m,t,d,ca,ctmp)
           tmp=1./(abs(ctmp)**2.)
           print*,i,tmp
 20     continue
      return
      end
c
c
      subroutine music(mm,m,ke,d,cr)
cccccccccccccccccccccccccccccccccccccccccccccccccccccccccccccccccccc
c This subroutine calculates MUSIC estimator.
c
c inputs
c      mm    dimension of cr(.) in main program
c      m     number of elements
c      ke    estimated number of sources
c      d(m)  element locations normalized by half wavelength
c      cr(m,m)    array output covariance matrix
c
c output    P(w) as a function of w
cccccccccccccccccccccccccccccccccccccccccccccccccccccccccccccccccccc
c
c global variables
      double precision d(mm)
      double complex cr(mm,mm)
c
c local variables
```

```fortran
      double precision pi,t,wk(480),tmp,lm(20)
      double complex ctmp,cz(20,20),ca(20),cf(210)
      pi=atan(1.d0)*4.
c
c eigenvalue decomposition
      do 10 i=1,m
      do 10 j=1,i
            cf(i*(i-1)/2+j)=cr(i,j)
 10   continue
      call eigch(cf,m,2,lm,cz,mm,wk,ier)
c
c calculate P(w)
      do 20 i=0,180
            t=float(i)*pi/180.
            tmp=0.
            do 30 j=1,m-ke
                do 40 im=1,m
                    ca(im)=cz(im,j)
 40             continue
                call atmuls(mm,m,t,d,ca,ctmp)
                tmp=tmp+abs(ctmp)**2.
 30         continue
            tmp=1./tmp
            print*,i,tmp
 20   continue
      return
      end
c
c
      subroutine strxxs(mm,m,a,d,cr,ca)
ccccccccccccccccccccccccccccccccccccccccccccccccccccccccccccccccccccc
c inputs
c     m       dimension of cr(.) in main program
c     m       dimension of matrix
c     a       angle
c     d       element locations
c     cr      Rxx (m x m matrix)
c
c output    ca      = st * Rxx * s
```

```fortran
ccccccccccccccccccccccccccccccccccccccccccccccccccccccccccccc
c
c global variable
      double precision a,d(mm)
      double complex ca,cr(mm,mm)
c
c local variable
      double precision tmp
c
      ca=cmplx(0.,0.)
      tmp=atan(1.d0)*4.*cos(a)
      do 10 i=1,m
      do 10 j=1,m
            ca=ca+cr(i,j)*cexp(cmplx(0.,(d(j)-d(i))*tmp))
  10    continue
      ca=ca/float(m)
      return
      end
c
c
      subroutine inverse(mm,m,cr,cinv,wk)
ccccccccccccccccccccccccccccccccccccccccccccccccccccccccccccccc
c inputs
c     mm   dimension of cr(.) and cinv(.) in main program
c     m      dimension of matrix
c     cr     m x m matrix
c     wk     working area (m)
c
c output     cinv   inverse of cr(.)
cccccccccccccccccccccccccccccccccccccccccccccccccccccccccccccc
c
c global variable
      double complex cr(mm,mm),cinv(mm,mm)
      double precision wk(mm)
c
c local variable
      double complex ctmp
c
      ctmp=cmplx(0.,0.)
```

```fortran
      do 10 i=1,m
          do 20 j=1,m
              cinv(i,j)=ctmp
20        continue
          cinv(i,i)=cmplx(1.,0.)
10    continue
      call leqt1c(cr,m,mm,cinv,m,mm,0,wk,ier)
      return
      end
c
c
      subroutine atmuls(mm,m,a,d,ca,cb)
ccccccccccccccccccccccccccccccccccccccccccccccccccccccccccccccccccc
c inputs
c     m       dimension of vector
c     a       angle
c     d       element locations
c     ca      complex vector
c
c output
c     cb      inner product of ca(.) and direction vector
ccccccccccccccccccccccccccccccccccccccccccccccccccccccccccccccccccc
c
c global variable
      double precision a,d(mm)
      double complex cb,ca(mm)
c
c local variable
      double precision tmp
c
      cb=cmplx(0.,0.)
      tmp=atan(1.d0)*4.*cos(a)
      do 10 i=1,m
          cb=cb+conjg(ca(i))*cexp(cmplx(0.,d(i)*tmp))
10    continue
      cb=cb/sqrt(float(m))
      return
      end
```

Problems

1. Derive (2.26) and show that $P_C(\omega_0) = M P + \sigma^2$ in a single target scene with ω_0 representing the angle of arrival. Further show that equality in (2.27) holds good only in a single target scene in presence of uncorrelated noise of identical variances, while pointing along the actual direction of arrival.

2. **Linear prediction method and maximum entropy method.** For any wide sense stationary stochastic process $x(n)$ with auto-correlation coefficients $r(k) = E[x(n+k)x^*(n)]$, $k = 0, \pm 1, \cdots, \pm \infty$, the power spectral density is given by

$$S_x(\omega) = \sum_{k=-\infty}^{\infty} r(k)e^{-jk\omega} \geq 0.$$

If the process is also regular (i.e., the set on which $S_x(\omega) = 0$ or ∞ is of measure zero) [37], then with $\mathbf{T}_k$ as defined below the associated entropy takes the form

$$H = \lim_{k \to \infty} \frac{\ln |\mathbf{T}_k|}{k+1} = \frac{1}{2\pi} \int_{-\pi}^{\pi} \ln S_x(\omega)d\omega \qquad (2.P.1)$$

and is well defined $(H > -\infty)$ [49].

a) For power spectral densities that are absolutely integrable over $|\omega| \leq \pi$, show that $S_x(\omega) \geq 0$ if and only if every hermitian Toeplitz matrix

$$\mathbf{T}_k = \begin{bmatrix} r(0) & r(1) & \cdots & r(k) \\ r^*(1) & r(0) & \cdots & r(k\text{-}1) \\ \vdots & \vdots & \ddots & \vdots \\ r^*(k) & r^*(k\text{-}1) & \cdots & r(0) \end{bmatrix}$$

is nonnegative definite, $k = 0, 1, \cdots, \infty$. Here

$$r(k) = \frac{1}{2\pi} \int_{-\pi}^{\pi} S_x(\omega)e^{jk\omega}d\omega, \quad k = 0, \pm 1, \cdots, \pm \infty.$$

b)   Subject to the additional constraint that in (2.P.1) the entropy $H > -\infty$, show that all $\mathbf{T}_k$ are positive definite, $k = 0, 1, \cdots, \infty$ [50].

c)   Given correlations $r(0), r(1), \cdots, r(M\text{-}1)$, the spectral extension problem is to extend this M correlations into a valid auto-correlation sequence such that the power spectral density generated from these correlations (given and extended) is nonnegative everywhere. If $|\mathbf{T}_{M-1}| > 0$, then show that the spectral extension problem has infinite number of solutions [51]. (If $|\mathbf{T}_{M-1}| = 0$, then the above extension problem has a unique solution.)

d)   Show that in c), the spectral extension subject to maximization of entropy (with respect to the free variables $r(M)$, $r(M+1)$, $\cdots$) leads to the system of equations represented by (2.33). Hence conclude that for one-variable processes, linear prediction method and maximum entropy method are equivalent. (This relationship was first exhibited by Van Den Bos [52]).

3.   Show that all zeros of the polynomial $g_{M-1}(z)$ in (2.41) lie outside the unit circle.

4.   [38, 41]. Derive the lower bounds in (2.145) and (2.146) by exploiting the nonnegative property of $G(\theta)$ in (2.15) together with the spatial distribution of sensor elements under consideration.

5.   Consider the scenario where the sensor noises are uncorrelated and identical with equal variance.

a)   In a single source scene, show that the signal subspace eigenvector coincides with the direction vector associated with the true arrival angle.

b)   In a two correlated source scene, derive the signal subspace eigenvalues and eigenvectors in terms of the signal parameters and their correlation coefficient.

6. **Effective correlation coefficient.**

    a)   Consider two equipowered perfectly coherent sources. Compute the effective signal correlation coefficient after employing i) the forward-only smoothing scheme once on the array, and ii) the f/b smoothing scheme once on the array.

    b)   Let $\rho_{ij} = |\rho_{ij}| e^{j\phi_{ij}}$ represent the correlation coefficient between signals $u_i(t)$ and $u_j(t)$ with respective arrival angles ω_i and ω_j. Show that, after applying the f/b smoothing scheme once on the array, the effective correlation coefficient between the same pair of signals becomes

$$\tilde{\rho}_{ij} = |\rho_{ij}| e^{-j(M-1)\omega_d} \cos((M-1)\omega_d + \phi_{ij}), \quad \omega_d = \frac{(\omega_i - \omega_j)}{2}.$$

Thus $|\tilde{\rho}_{ij}| \leq |\rho_{ij}|$, i.e., the f/b scheme has in effect decorrelated the signals beyond their original correlation level. Since uncorrelated signals have superior performance in this new scheme (refer to the discussion after (3.83)), any amount of decorrelation will essentially lead to improved performance.

References

[1]   H. L. Van Trees, *Detection, Estimation, and Modulation Theory*, Part I. New York: John Wiley and Sons, 1968.

[2]   R. A. Monzingo and T. W. Miller, *Introduction to Adaptive Arrays*. New York: John Wiley and Sons, 1980.

[3]   T. G. Kincaid, "The complex representation of signals," *General Electric Company Report*, No. R67EMH5, Oct. 1982.

[4]   A. Papoulis, *Probability, Random Variables and Stochastic Processes*, 2nd ed. New York: McGraw-Hill, 1984.

[5]   Y. T. Lo, "A mathematical theory of antenna arrays with randomly placed elements," *IEEE Trans. Antennas Propagat.*, vol. AP-12, pp. 257-268, May 1964.

[6]   B. D. Steinberg, *Principles of Aperture and Array System Design*. New York: John Wiley and Sons, 1976.

[7] M. S. Bartlett, "Periodogram analysis and continuous spectra," *Biometrica*, vol. 37, pp. 1-16, 1950.

[8] G. Jenkins and D. Watts, *Spectral Analysis and Its Applications*. San Francisco, CA: Holden-Day, 1968.

[9] J. Capon, "High resolution frequency-wavenumber spectrum analysis," *Proc. IEEE*, vol. 57, pp. 1408-1418, Aug. 1969.

[10] S. W. Lang and J. H. McClellan, "Frequency estimation with maximum entropy spectral estimators," *IEEE Trans. Acoust., Speech, Signal Processing*, vol. ASSP-28, pp. 716-724, Dec. 1980.

[11] R. N. McDonough, "Application of the maximum-likelihood method and the maximum-entropy method to array processing," in *Nonlinear Methods of Spectral Analysis*. S. Haykin, Ed. New York: Springer-Verlag, 1983.

[12] S. Haykin, *Adaptive Filter Theory*. Englewood Cliffs, NJ: Prentice-Hall, 1986.

[13] M. Bôcher, *Introduction to Higher Algebra*. New York: Macmillan, 1907.

[14] R. T. Lacoss, "Data adaptive spectral analysis methods," *Geophysics*, vol. 36, pp. 661-675, Aug. 1971.

[15] D. P. Robbins and H. Ru, "Determinants and alternating sign matrices," *Advances in Mathematics*, vol. 62, pp. 169-184, 1986.

[16] C. L. Dodgson, "Condensation of determinants," *Proc. Royal Soc. London*, vol. 15, pp. 150-155, 1866.

[17] J. P. Burg, "The relationship between maximum entropy spectra and maximum likelihood spectra," *Geophysics*, vol. 37, pp. 375-376, Apr. 1972.

[18] V. F. Pisarenko, "The retrieval of harmonics from a covariance function," *Geophys. J. Royal Astron. Soc.*, vol. 33, pp. 347-366, 1973.

[19] R. O. Schmidt, "Multiple emitter location and signal parameter estimation," in *Proc. RADC Spectral Est. Workshop*, Oct. 1979, pp. 243-258.

[20] G. Bienvenu and L. Kopp, "Adaptivity to background noise spatial coherence for high resolution passive methods," in *Proc. IEEE ICASSP '80*, Denver, CO, pp. 307-310.

[21] B. Widrow, K. M. Duvall, R. P. Gooch, and W. C. Newman, "Signal cancellation phenomena in adaptive antennas: Causes and cures," *IEEE Trans. Antennas Propagat.*, vol. AP-30, pp. 469-478, May 1982.

[22] W. F. Gabriel, "Spectral analysis and adaptive array superresolution techniques," *Proc. IEEE*, vol. 68, pp. 654-666, June 1980.

[23] ---, "Adaptive superresolution of coherent RF spatial sources," in *Proc. First ASSP Workshop Spectral Estimation*, Communication Research Lab., McMaster University, Aug. 1981.

[24] J. E. Evans, J. R. Johnson, and D. F. Sun, "High resolution angular spectrum estimation techniques for terrain scattering analysis and angle of arrival estimation," in *Proc. First ASSP Workshop Spectral Estimation*, Communication Research Lab., McMaster University, Aug. 1981.

[25] ---, "Application of advanced signal processing techniques to angle of arrival estimation in ATC navigation and surveillance system," *M.I.T. Lincoln Lab.*, Lexington, MA, Rep. 582, 1982.

[26] T. J. Shan, M. Wax, and T. Kailath, "On spatial smoothing for estimation of coherent signals," *IEEE Trans. Acoust., Speech, Signal Processing*, vol. ASSP-33, pp. 806-811, Aug. 1985.

[27] F. Haber and M. Zoltowski, "Spatial spectrum estimation in a coherent signal environment using an array in motion," *IEEE Trans. on Ant. and Propagat.*, vol. AP-34, pp. 301-310, Mar. 1986.

[28] C. N. Dorny, *A Vector Space Approach to Models and Optimization*. New York: John Wiley and Sons, 1975.

[29] D. C. Youla, *Private Communication*, May 1987.

[30] S. U. Pillai, Y. Barness, and F. Haber, "A new approach to array geometry for improved spatial spectrum estimation," *Proc. IEEE*, vol. 73, pp. 1522-1524, Oct. 1985.

[31] S. Haykin, "Radar Array Processing for angle of arrival estimation," in *Array Signal Processing*, S. Haykin, Ed. Englewood Cliffs, NJ: Prentice-Hall, 1985, pp. 194-289.

[32] T. J. Shan and T. Kailath, "Adaptive beamforming for coherent signals and interference," *IEEE Trans. Acoust., Speech, Signal Processing*, vol. ASSP-33, pp. 527-536, June 1985.

[33] S. U. Pillai and B. H. Kwon, "An improved spatial smoothing technique for coherent signal identification," in *ONR Annual Report*, Polytechnic University, June 1987.

[34] Y. Bresler and A. Macovski, "On the number of signals resolvable by a uniform linear array," *IEEE Trans. Acoust., Speech, Signal Processing*, vol. ASSP-34, pp. 1361-1375, Dec. 1986.

[35] F. R. Gantmacher, *The Theory of Matrices*, vol. 1. New York: Chelsea, 1977.

[36] T. W. Anderson, *An Introduction to Multivariate Statistical Analysis*, 2nd Ed. New York: John Wiley and Sons, 1984.

[37] U. Grenander and G. Szegö, *Toeplitz Forms and Their Applications*. New York: Chelsea, 1984.

[38] L. Rédei and A. Rényi, "On the representation of 1, 2, $\cdots$, n by differences," *Mat. Sbornik (Recueil Math.)*, vol. 66 (NS 24), pp. 385-389 (Russian), 1949.

[39] A. Brauer, "A problem of additive number theory and its application in electrical engineering," *J. Elisha Mitchell Sci. Soc.*, vol. 61, pp. 55-66, 1945.

[40] P. Erdös and I. S. Gál, "On the representation of 1, 2, $\cdots$, N by differences," *Nederl. Akad. Wetensch. Proc.*, vol. 51, pp. 1155-1158, 1948; *Indagationes Math.*, vol. 10, pp. 379-382, 1948.

[41] J. Leech, "On the representation of 1, 2, $\cdots$, n by differences," *J. London Math. Soc.*, vol 31, pp. 160-169, 1956.

[42] D. Pearson, S. U. Pillai, and Y. Lee, "An algorithm for near-optimal placement of sensor elements," in *ONR Annual Report*, Polytechnic University, June 1988.

[43] A. Paulraj, R. Roy, and T. Kailath, "Estimation of signal parameters via rotational invariance techniques - ESPRIT," in *Proc. 19th Asilomar Conf.*, Pacific Grove, CA, Nov. 1985.

[44] R. Roy, A. Paulraj, and T. Kailath, "ESPRIT - a subspace rotation approach to estimation of parameters of cisoids in noise," *IEEE Trans. Acoust., Speech, Signal Processing*, vol. ASSP-34, pp. 1340-1342, Oct. 1986.

[45] R. H. Roy, "ESPRIT : Estimation of Signal Parameter via Rotational Invariance Technique," Ph.D. thesis, Stanford Univ., Stanford, CA, 1987.

[46] S. U. Pillai and B. H. Kwon, "GEESE (GEneralized Eigenvalues utilizing Signal subspace Eigenvectors) - A new technique for direction finding," *Proc. Twenty Second Annual Asilomar Conference on Signals, Systems, and Computers*, Pacific Grove, CA, Oct. 31 - Nov. 2, 1988.

[47] B. H. Kwon, "New high resolution techniques and their performance analysis for angle-of-arrival estimation," Ph.D. dissertation, Polytechnic Univ., Brooklyn, NY, 1989.

[48] Y. Lee, S. U. Pillai and D. C. Youla, "Direction finding from first order statistics," in *ONR Annual Report*, Polytechnic University, Dec. 1987.

[49] G. Szegö, *Collected Papers*, vol. 3, R. Askey ed. Boston: Birkhäuser, 1982.

[50] D. C. Youla and N. N. Kazanjian, "Bauer-type factorization of positive matrices and the theory of matrix polynomials orthogonal on the unit circle," *IEEE Trans. Circuits Systs.*, vol. CAS-25, pp. 57-69, Feb. 1978.

[51] D. C. Youla, "The FEE: A new tunable high-resolution spectral estimator," Part I, Technical note no. 3, Department of Electrical Engineering, Polytechnic Institute of New York, Brooklyn, New York, 1980.

[52] A. Van Den Bos, "Alternative interpretation of maximum entropy spectral analysis," *IEEE Trans. Inform. Theory*, vol. IT-17, pp. 493-494, July 1971.

Chapter 3
Performance Analysis

3.1 Introduction

So far we have assumed that exact ensemble averages of the array output covariances (or mean values) are available, and based on this assumption, several conventional and high resolution techniques have been developed to resolve the directions of arrival of incoming signals. In this chapter we analyze the performance of these methods based on finite observations, from a statistical point of view, and establish several results. When the ensemble averages are not available, they usually are estimated from the available array output data. In general, a finite data sample is available and estimation is carried out for the unknown covariances of interest such that the resulting estimators represent their "most likely" values. The principle of maximum likelihood (ML) is often chosen in various estimation and hypothesis testing problems for this purpose [1]. As the name implies, the unknowns are selected so as to maximize (the logarithm of) the likelihood function, which is (the logarithm of) the joint probability density function of observations.

For parametric estimation problems, the desirable properties of estimators corresponding to the unknown parameters of interest, are unbiasedness and low values for their variance. For estimators of this type, general results are known regarding their minimum attainable variance. The Fisher-Cramer-Rao (generally known as Cramer-Rao or C-R) bound gives the absolute lower bound for the variance of any unbiased estimator satisfying regularity conditions [2, 3]. Estimators that achieve the C-R bound are said to be efficient. For large data size, the ML estimates are asymptotically efficient or consistent. In addition, these estimates possess the property of invariance, which states that if on the basis of a given sample, $\hat{\theta}_1, \hat{\theta}_2, \cdots, \hat{\theta}_K$ are the ML estimates of the parameters $\theta_1, \theta_2, \cdots, \theta_K$, then $\phi_i(\hat{\theta}_1, \cdots, \hat{\theta}_K)$, $i =,$ $2, \cdots, K$ are the ML estimates of $\phi_1, \phi_2, \cdots$ if the transformations $\phi_1, \phi_2, \cdots, \phi_K$ to $\theta_1, \theta_2, \cdots, \theta_K$ are one-to-one [4].

In the absence of such a 1-1 transformation, often it is difficult to obtain unbiased estimators for the unknown parameters of interest, especially in a multiple parameter case. Under those circumstances, one usually turns to indirect means of estimating the unknown parameters. In the array processing problem, the unknowns of interest are, for example, the angles of arrival and the power levels of the multiple signals. Techniques described in the previous chapter are examples that realize this goal indirectly; angles of arrival in these cases correspond to the locations of predominant peaks (or zeros) in their outputs. When estimates of covariances obtained from the data are used instead in computing these outputs, one hopes that they will also be consistent, in turn guaranteeing accurate estimation of the unknown parameters of interest. These asymptotic properties are established here for the conventional and high resolution estimators. Finally, the question of resolution of closely spaced sources is addressed and resolution thresholds are derived for two source scenes in uncorrelated and coherent situations.

3.2 The Maximum Likelihood Estimate of the Covariance Matrix and Some Related Distributions

Let

$$\mathbf{x}(n) \triangleq \mathbf{x}(t_n) = \left[x_1(t_n), x_2(t_n), \cdots, x_M(t_n)\right]^T \tag{3.1}$$

represent an M-element array output vector at time instant t_n (as in (2.19) or (2.51)) and further let $\mathbf{x}(n)$, $n = 1, 2, \cdots, N$ denote an independent sample of N such complex valued output vectors. We will assume that these outputs are M-variate zero mean complex Gaussian random vectors with unknown covariance matrix,

$$\mathbf{R} = E[\mathbf{x}(t_n)\mathbf{x}^\dagger(t_n)]. \tag{3.2}$$

Thus, the probability density function (p.d.f) of $\mathbf{x}(n)$ is given by [5]

$$f_\mathbf{x}(\mathbf{x}(n)) = \frac{1}{|\pi\mathbf{R}|}e^{-\mathbf{x}^\dagger(n)\mathbf{R}^{-1}\mathbf{x}(n)}$$

and hence the likelihood function takes the form

$$L \triangleq f(\mathbf{x}(1), \mathbf{x}(2), \cdots, \mathbf{x}(N)) = \prod_{n=1}^{N} f_{\mathbf{x}}(\mathbf{x}(n))$$

$$= \frac{1}{|\pi \mathbf{R}|^N} e^{-\sum_{n=1}^{N} \mathbf{x}^\dagger(n)\mathbf{R}^{-1}\mathbf{x}(n)} = \frac{1}{|\pi \mathbf{R}|^N} e^{-tr(\mathbf{R}^{-1}\mathbf{B})} \tag{3.3}$$

where

$$\mathbf{B} = \sum_{n=1}^{N} \mathbf{x}(n)\mathbf{x}^\dagger(n). \tag{3.4}$$

This gives the log-likelihood function to be

$$\log L = -MN \ln \pi - N \ln |\mathbf{R}| - tr(\mathbf{R}^{-1}\mathbf{B}). \tag{3.5}$$

For positive definite $\mathbf{B}$, to determine the maximum of the above likelihood function with respect to $\mathbf{R}$, let $\mathbf{B} = \mathbf{DD}^\dagger$ and define $\mathbf{D}^\dagger \mathbf{R}^{-1}\mathbf{D} = \mathbf{E}$. Then $\mathbf{R} = \mathbf{DE}^{-1}\mathbf{D}^\dagger$ and $|\mathbf{R}| = |\mathbf{D}| |\mathbf{E}^{-1}| |\mathbf{D}^\dagger| = |\mathbf{B}|/|\mathbf{E}|$. Thus $tr(\mathbf{R}^{-1}\mathbf{B}) = tr(\mathbf{R}^{-1}\mathbf{DD}^\dagger) = tr(\mathbf{D}^\dagger\mathbf{R}^{-1}\mathbf{D}) = tr(\mathbf{E})$ and from (3.5)

$$\Lambda(\mathbf{R}) \triangleq -N \ln |\mathbf{R}| - tr(\mathbf{R}^{-1}\mathbf{B})$$

$$= -N \ln |\mathbf{B}| + N \ln |\mathbf{E}| - tr(\mathbf{E}). \tag{3.6}$$

Since $\mathbf{E}$ is also positive definite, it can be factorized as $\mathbf{E} = \mathbf{LL}^\dagger$, where $\mathbf{L}$ is lower triangular with positive diagonal elements ($L_{ii} > 0$) and

$$\Lambda(\mathbf{R}) = -N \ln |\mathbf{B}| - \sum_{i=1}^{M} (N \ln L_{ii}^2 - L_{ii}^2) - \sum_{i>j} |L_{ij}|^2$$

is maximized for $L_{ii}^2 = N$ and $L_{ij} = 0$, $i \neq j$; i.e., $\mathbf{E} = N\mathbf{I}$ or the maximum of the likelihood function (3.5) with respect to the positive definite matrix $\mathbf{R}$ occurs at

$$\hat{\mathbf{R}} = \frac{1}{N}\mathbf{DD}^\dagger = \frac{1}{N}\mathbf{B} = \frac{1}{N}\sum_{n=1}^{N} \mathbf{x}(n)\mathbf{x}^\dagger(n) \triangleq \mathbf{S}. \tag{3.7}$$

$\hat{\mathbf{R}}$ is a hermitian random matrix and the joint p.d.f. for the elements

B_{ij}, $i \leq j = 1, 2, ..., M$ of $\mathbf{B} \triangleq N\hat{\mathbf{R}}$ is known as the complex Wishart distribution of order N and is denoted by $CW(N, M, \mathbf{R})$. Goodman has shown that distribution to be [5]

$$f_B(\mathbf{B}) = \frac{|\mathbf{B}|^{N-M}}{\Gamma_M(N) |\mathbf{R}|^N} e^{-tr(\mathbf{R}^{-1}\mathbf{B})} \tag{3.8}$$

where

$$\Gamma_M(N) = \pi^{M(M-1)/2} \Gamma(N) \Gamma(N-1) \cdots \Gamma(N-M+1).$$

(See Appendix 3.A for some related distributions and results.) These ML estimates of covariances in (3.7) can be used in computations whenever their exact counterparts are unknown. For example, the estimate of the beamformer $P_B(\omega)$ is given by

$$\hat{P}_B(\omega) = \mathbf{a}^\dagger(\omega) \hat{\mathbf{R}} \mathbf{a}(\omega). \tag{3.9}$$

Using (3.7)

$$\hat{P}_B(\omega)/P_B(\omega) = \frac{1}{N} \sum_{n=1}^{N} |\mathbf{a}^\dagger(\omega)\mathbf{x}(n)|^2/P_B(\omega). \tag{3.10}$$

However,

$$|\mathbf{a}^\dagger(\omega)\mathbf{x}(n)|^2/P_B(\omega) = \left| \frac{\mathbf{a}^\dagger(\omega)\mathbf{x}(n)}{(\mathbf{a}^\dagger(\omega)\mathbf{R}\mathbf{a}(\omega))^{1/2}} \right|^2 \triangleq |y(n)|^2. \tag{3.11}$$

From the joint normality of $\mathbf{x}(n)$, note that $y(n)$ is also normal with

$$E[y(n)] = 0 \quad \text{and} \quad E[|y(n)|^2] = 1.$$

Thus, $y(n) \sim N(0, 1)$ and $|y(n)|^2$ is a χ^2 random variable with one degree of freedom; i.e., $|y(n)|^2 \sim \chi^2(1)$. Further, from the independence of data samples, $y(n)$, $n = 1, 2, ..., N$ are also independent and hence

$$\sum_{n=1}^{N} |y(n)|^2 \sim \chi^2(N).$$

This together with (3.10) and (3.11) gives

$$\frac{N\hat{P}_B(\omega)}{P_B(\omega)} \sim \chi^2(N). \tag{3.12}$$

Since a χ^2 random variable has mean and variance equal to its degree of freedom [2], from (3.12) we have

$$E[\hat{P}_B(\omega)] = P_B(\omega)$$

and

$$Var[\hat{P}_B(\omega)] = \frac{1}{N}(P_B(\omega))^2 \to 0 \quad as \quad N \to \infty$$

implying consistency for the estimator in (3.9).

Similarly, the estimator for Capon's minimum variance power output $P_C(\omega)$ in (2.26) is given by

$$\hat{P}_C(\omega) = \frac{1}{\mathbf{a}^\dagger(\omega)\hat{\mathbf{R}}^{-1}\mathbf{a}(\omega)}. \tag{3.13}$$

Using (3.A.20) and (3.7),

$$N\frac{\mathbf{a}^\dagger(\omega)\mathbf{R}^{-1}\mathbf{a}(\omega)}{\mathbf{a}^\dagger(\omega)\hat{\mathbf{R}}^{-1}\mathbf{a}(\omega)} = \frac{N\hat{P}_C(\omega)}{P_C(\omega)} \sim \chi^2(N-M+1). \tag{3.14}$$

This also gives

$$E[\hat{P}_C(\omega)] = \left(\frac{N-M+1}{N}\right)P_C(\omega)$$

and

$$Var[\hat{P}_C(\omega)] = \left(\frac{N-M+1}{N^2}\right)(P_C(\omega))^2 \to 0 \quad as \quad N \to \infty.$$

Thus, asymptotically, $\hat{P}_C(\omega)$ is unbiased and consistent.

To analyze the performance of the estimator corresponding to the linear prediction method, it is best to make use of $P_L(\omega)$ given by (2.46). In that case,

$$\hat{P}_L(\omega) = \frac{\hat{\delta}_{M-1}}{|\hat{H}(e^{j\omega})|^2} \tag{3.15}$$

where

$$\hat{\delta}_{M-1} = \hat{r}(0) - \hat{\gamma}_{M-1,1}^{\dagger}\hat{T}_{M-2}^{-1}\hat{\gamma}_{M-1,1} \tag{3.16}$$

and

$$\hat{H}(e^{j\omega}) = 1 - s^{\dagger}(\omega)\hat{T}_{M-2}^{-1}\hat{\gamma}_{M-1,1}. \tag{3.17}$$

In the case of an M-element uniform array from (2.36),

$$\mathbf{R} \triangleq \mathbf{T}_{M-1} = \begin{bmatrix} \mathbf{T}_{M-2} & \gamma_{M-1,1} \\ \\ \gamma_{M-1,1}^{\dagger} & r(0) \end{bmatrix}$$

and hence

$$N\hat{\mathbf{R}} = N\hat{\mathbf{T}}_{M-1} = N\begin{bmatrix} \hat{\mathbf{T}}_{M-2} & \hat{\gamma}_{M-1,1} \\ \\ \hat{\gamma}_{M-1,1}^{\dagger} & \hat{r}(0) \end{bmatrix} \sim CW(N,M,\mathbf{R}).$$

Using (3.A.8), the lower right-hand corner 1×1 Schur complement $N\hat{\delta}_{M-1}$ given by (3.16) has the distribution $CW(N-M+1,1,\delta_{M-1})$ or

$$\frac{N\hat{\delta}_{M-1}}{\delta_{M-1}} \sim \chi^2(N-M+1). \tag{3.18}$$

Similarly, using (3.A.9) and (3.A.15), the regression coefficient $\hat{T}_{M-2}^{-1}\hat{\gamma}_{M-1,1}$ is conditionally Gaussian when $\hat{T}_{M-2}$ is given and more importantly, $\hat{\delta}_{M-1}$ is statistically independent of the regression coefficient $\hat{T}_{M-2}^{\dagger}\hat{\gamma}_{M-1,1}$ and $\hat{T}_{M-2}$. Thus, $\hat{H}(e^{j\omega})$ is statistically independent of $\hat{\delta}_{M-1}$ and with

$$\frac{N\hat{P}_L(\omega)}{P_L(\omega)} = \frac{N\hat{\delta}_{M-1}/\delta_{M-1}}{|\hat{H}(e^{j\omega})/H(e^{j\omega})|^2} = \frac{X}{T^2} \tag{3.19}$$

we also have

$$X = N \, \hat{\delta}_{M-1}/\delta_{M-1} \;\sim\; \chi^2(N-M+1).$$

Clearly, the numerator and denominator in (3.19) are statistically independent [6].

3.3 Performance Analysis of Covariance Based Eigenvector Techniques : MUSIC and Spatial Smoothing Schemes

In this section, the statistical properties of eigenvector-based, forward-only and forward/backward high resolution smoothing schemes described in section 2.3 are examined in detail. In particular, by employing an asymptotic analysis [4, 7], first order approximations to the mean and variance of the null estimator $\tilde{Q}(\omega)$ in (2.140) corresponding to the forward/backward smoothing scheme is presented. Similar results for the forward-only scheme and the standard MUSIC scheme are obtained as special cases of this general analysis. Finally, using these results resolution thresholds for two closely spaced equipowered sources in both uncorrelated and coherent scenes are derived.

To begin with, it is assumed that the signals and noises are stationary, zero mean circular Gaussian[1] independent random processes, and further, the noises are assumed to be independent and identical between themselves with common variance σ^2. Thus, the array output observations $\mathbf{x}(n)$, $n = 1, 2, \cdots, N$, represent, zero mean M-variate circular Gaussian, independent samples with unknown covariance matrix $\mathbf{R}$. The ML estimate $\mathbf{S}$ of this covariance matrix is given by (3.7). Using the invariant property of the maximum likelihood procedure, the corresponding estimates $\mathbf{S}_f$, $\mathbf{S}_b$ and $\tilde{\mathbf{S}}$ for the unknown smoothed matrices $\mathbf{R}^f$, $\mathbf{R}^b$ and $\tilde{\mathbf{R}}$ in (2.112), (2.120) and

---

(1) A complex random vector $\mathbf{z}$ is defined to be circular Gaussian if its real part $\mathbf{x}$ and imaginary part $\mathbf{y}$ are jointly Gaussian and their joint covariance matrix has the form [8, 9]

$$E\left[\begin{bmatrix}\mathbf{x}\\\mathbf{y}\end{bmatrix}\begin{bmatrix}\mathbf{x}^T & \mathbf{y}^T\end{bmatrix}\right] = \frac{1}{2}\begin{bmatrix}\mathbf{V} & -\mathbf{W}\\\mathbf{W} & \mathbf{V}\end{bmatrix}$$

where $\mathbf{z} = \mathbf{x} + j\,\mathbf{y}$. When $\mathbf{z}$ has zero mean, its covariance matrix is given by $E[\mathbf{z}\mathbf{z}^\dagger] \triangleq E\left[(\mathbf{x}+j\mathbf{y})(\mathbf{x}^T - j\mathbf{y}^T)\right] = \mathbf{V} + j\mathbf{W}$. Clearly, $E\left(\mathbf{z}\mathbf{z}^T\right) = \mathbf{O}$.

(2.127) respectively, can be constructed from S by the same rule that is used in constructing R^f, R^b and $\tilde{R}$, respectively, from R. Thus, for example,

$$\tilde{S} = \frac{1}{2}(S_f + S_b) \tag{3.20}$$

and

$$S_f = \frac{1}{L} \sum_{l=1}^{L} S_l^f = \frac{1}{NL} \sum_{l=1}^{L} \sum_{n=1}^{N} x_l^f(n)(x_l^f(n))^\dagger \tag{3.21}$$

etc. Though the random matrix NS and $NS_l^f, l = 1, 2, \cdots, L$ have central complex Wishart distributions [5], the p.d.f of the estimated smoothed matrices in (3.20) − (3.21) is not a complex Wishart distribution. In what follows we study the statistical properties of these estimated smoothed covariance matrices and their associated sample estimators for direction finding.

3.3.1 Asymptotic Distribution of Eigenparameters Associated with Smoothed Sample Covariance Matrices

With $\tilde{S}$ representing the ML estimate of the forward/backward (f/b) smoothed covariance matrix $\tilde{R}$, we have

$$\tilde{R} = \frac{1}{2L} \sum_{l=1}^{L} \left(R_l^f + R_l^b \right) \triangleq \tilde{B}\tilde{A}\tilde{B}^\dagger, \tag{3.22}$$

$$\tilde{S} = \frac{1}{2L} \sum_{l=1}^{L} \left[\frac{1}{N} \sum_{n=1}^{N} \left(x_l^f(n)(x_l^f(n))^\dagger + x_l^b(n)(x_l^b(n))^\dagger \right) \right] = \tilde{E}\tilde{L}\tilde{E}^\dagger$$

where

$$E[\tilde{S}] = \tilde{R}$$

$$\tilde{B} = \left[\tilde{\beta}_1, \tilde{\beta}_2, \cdots, \tilde{\beta}_M \right], \quad \tilde{E} = \left[\tilde{e}_1, \tilde{e}_2, \cdots, \tilde{e}_M \right]$$

$$\tilde{A} = diag[\tilde{\lambda}_1, \tilde{\lambda}_2, \cdots, \tilde{\lambda}_K, \sigma^2, \cdots, \sigma^2]$$

$$\tilde{L} = diag[\tilde{l}_1, \tilde{l}_2, \cdots, \tilde{l}_K, \tilde{l}_{K+1}, \cdots, \tilde{l}_M]$$

$$\tilde{\mathbf{B}}\,\tilde{\mathbf{B}}^{\dagger} = \tilde{\mathbf{E}}\,\tilde{\mathbf{E}}^{\dagger} = \mathbf{I}_M$$

and $\tilde{\mathbf{B}}$, $\tilde{\mathbf{E}}$ satisfies $\tilde{\beta}_{ii}$, $\tilde{e}_{ii} \geq 0$, $i = 1,,2,\cdots,M$ for uniqueness. Here the normalized vectors $\tilde{\mathbf{e}}_1$, $\tilde{\mathbf{e}}_2$, $\cdots$, $\tilde{\mathbf{e}}_K$ are the ML estimates of the normalized eigenvectors $\tilde{\boldsymbol{\beta}}_1$, $\tilde{\boldsymbol{\beta}}_2$, $\cdots$, $\tilde{\boldsymbol{\beta}}_K$ of $\tilde{\mathbf{R}}$ respectively. Similarly, $\tilde{l}_1$, $\tilde{l}_2$, $\cdots$, $\tilde{l}_K$ are the ML estimates of the K largest and distinct eigenvalues $\tilde{\lambda}_1$, $\tilde{\lambda}_2$, $\cdots$, $\tilde{\lambda}_K$ and the mean of $\tilde{l}_{K+1}$, $\cdots$, $\tilde{l}_M$ represent the sample estimate for the repeating lowest eigenvalue σ^2 of $\tilde{\mathbf{R}}$. Following (2.140), the sample direction estimator can be written as

$$\hat{Q}(\omega) = \sum_{k=K+1}^{M} |\tilde{\mathbf{e}}_i^{\dagger}\mathbf{a}(\omega)|^2 = 1 - \sum_{k=1}^{K} |\tilde{\mathbf{e}}_i^{\dagger}\mathbf{a}(\omega)|^2. \qquad (3.23)$$

As is well known (see Appendix 3.B), the eigenvectors are not unique, and let $\mathbf{C}$ represent yet another set of eigenvectors for $\tilde{\mathbf{S}}$, i.e.,

$$\tilde{\mathbf{S}} = \mathbf{C}\,\tilde{\mathbf{L}}\,\mathbf{C}^{\dagger} \qquad (3.24)$$

where

$$\mathbf{C} = \left[\mathbf{c}_1, \mathbf{c}_2, \cdots, \mathbf{c}_M\right] \qquad (3.25)$$

and

$$\mathbf{C}\,\mathbf{C}^{\dagger} = \mathbf{I}_M. \qquad (3.26)$$

For reasons that will become apparent later, $\mathbf{C}$ is made unique here by requiring that all diagonal elements of $\mathbf{Y}$ given by

$$\mathbf{Y} = \tilde{\mathbf{B}}^{\dagger}\mathbf{C} \qquad (3.27)$$

are positive ($y_{ii} \geq 0$, $i = 1, 2, \cdots, M$). In what follows, we first derive the asymptotic distribution of the set of sample eigenvectors and eigenvalues of $\tilde{\mathbf{S}}$ given by (3.24) – (3.27) and use this to analyze the performance of the sample directions of arrival estimator $\hat{Q}(\omega)$ in (3.23). This is made possible by noticing that although the estimated eigenvectors $\tilde{\mathbf{e}}_1$, $\tilde{\mathbf{e}}_2$, $\cdots$, $\tilde{\mathbf{e}}_K$, in (3.23) are structurally identical to their counterparts $\tilde{\boldsymbol{\beta}}_1$, $\tilde{\boldsymbol{\beta}}_2$, $\cdots$, $\tilde{\boldsymbol{\beta}}_K$, and in particular have $\tilde{e}_{ii} \geq 0$, nevertheless as shown in Appendix 3.B, they are also related to $\mathbf{c}_i$, $i = 1, 2, \cdots, K$, through a phase factor; i.e.,

- 117 -

$$\tilde{\mathbf{e}}_i = e^{j\phi_i} \mathbf{c}_i, \quad i = 1, 2, \cdots, K \tag{3.28}$$

and hence

$$\hat{Q}(\omega) = 1 - \sum_{i=1}^{K} |\tilde{\mathbf{e}}_i^\dagger \mathbf{a}(\omega)|^2 = 1 - \sum_{i=1}^{K} |\mathbf{c}_i^\dagger \mathbf{a}(\omega)|^2$$

$$= 1 - \sum_{i=1}^{K} \hat{y}_i(\omega) \tag{3.29}$$

with

$$\hat{y}_i(\omega) = |\mathbf{c}_i^\dagger \mathbf{a}(\omega)|^2. \tag{3.30}$$

Thus, the statistical properties of $\hat{Q}(\omega)$ can be completely specified by those of $\mathbf{c}_i$, $i = 1, 2, \cdots, K$, and toward this purpose, let

$$\mathbf{F} = \sqrt{N}\,(\tilde{\mathbf{L}} - \tilde{\mathbf{A}}) \tag{3.31}$$

$$\mathbf{G} = [\mathbf{g}_1, \mathbf{g}_2, \cdots, \mathbf{g}_K, \cdots, \mathbf{g}_M] = \sqrt{N}\,(\mathbf{C} - \tilde{\mathbf{B}}) \tag{3.32}$$

and

$$\mathbf{T} = \tilde{\mathbf{B}}^\dagger \tilde{\mathbf{S}} \tilde{\mathbf{B}} = \tilde{\mathbf{B}}^\dagger \mathbf{C} \tilde{\mathbf{L}} \mathbf{C}^\dagger \tilde{\mathbf{B}} = \mathbf{Y} \tilde{\mathbf{L}} \mathbf{Y}^\dagger \tag{3.33}$$

where $\mathbf{Y}$ is as defined in (3.27) with $y_{ii} \geq 0$, $i = 1, 2, \cdots, M$. Further, let

$$\mathbf{U} = \sqrt{N}\,(\mathbf{T} - \tilde{\mathbf{A}})$$

$$= \frac{1}{\sqrt{N}} \sum_{n=1}^{N} \left[\frac{1}{2L} \sum_{l=1}^{L} \tilde{\mathbf{B}}^\dagger (\mathbf{x}_l^f(n)(\mathbf{x}_l^f(n))^\dagger + \mathbf{x}_l^b(n)(\mathbf{x}_l^b(n))^\dagger) \tilde{\mathbf{B}} - \tilde{\mathbf{A}} \right]$$

$$= \frac{1}{\sqrt{N}} \sum_{n=1}^{N} \left[\frac{1}{2L} \sum_{l=1}^{L} (\mathbf{z}_l^f(n)(\mathbf{z}_l^f(n))^\dagger + \mathbf{z}_l^b(n)(\mathbf{z}_l^b(n))^\dagger) - \tilde{\mathbf{A}} \right] \tag{3.34}$$

with

$$\mathbf{z}_l^f(n) = \tilde{\mathbf{B}}^\dagger \mathbf{x}_l^f(n) \sim N(0, \tilde{\mathbf{B}}^\dagger \mathbf{R}_l^f \tilde{\mathbf{B}}) \tag{3.35}$$

and

$$\mathbf{z}_l^b(n) = \tilde{\mathbf{B}}^\dagger \mathbf{x}_l^b(n) \sim N(0, \tilde{\mathbf{B}}^\dagger \mathbf{R}_l^b \tilde{\mathbf{B}}). \tag{3.36}$$

- 118 -

It is easily verified that these random vectors preserve the circular Gaussian property of the data vectors. Again, from the independence of observations, asymptotically every entry in **U** is a sum of a large number of independent random variables. Using the multivariate central limit theorem [4], the limiting distribution of **U** tends to be normal with means and covariances given by

$$E[u_{ij}] = \frac{1}{\sqrt{N}} \sum_{n=1}^{N} E\left[\frac{1}{2L} \sum_{l=1}^{L} \left(z_{li}^{f}(n)z_{lj}^{f\,*}(n) + z_{li}^{b}(n)z_{lj}^{b\,*}(n)\right) - \tilde{\lambda}_i\,\delta_{ij}\right]$$

$$= 0 \tag{3.37}$$

(From here onwards, whenever there is no confusion, we will suppress the time index n) since

$$E\left[\frac{1}{2L} \sum_{l=1}^{L} (z_{li}^{f}(z_{lj}^{f})^{*} + z_{li}^{b}(z_{lj}^{b})^{*})\right]$$

$$= \tilde{\beta}_i^{\dagger} E\left[\frac{1}{2L} \sum_{l=1}^{L} \left(\mathbf{x}_l^{f}(\mathbf{x}_l^{f})^{\dagger} + \mathbf{x}_l^{b}(\mathbf{x}_l^{b})^{\dagger}\right)\right]\tilde{\beta}_j$$

$$= \tilde{\beta}_i^{\dagger} E[\tilde{\mathbf{S}}]\tilde{\beta}_j = \tilde{\beta}_i^{\dagger}\tilde{\mathbf{R}}\tilde{\beta}_j = \tilde{\lambda}_i\,\delta_{ij} \tag{3.38}$$

and

$$E[u_{ij}u_{kl}^{*}] = \frac{1}{N} \sum_{n=1}^{N} \left[\frac{1}{4L^2} \sum_{p=1}^{L} \sum_{q=1}^{L} \left\{ E[z_{pi}^{f} z_{pj}^{f\,*} z_{qk}^{f\,*} z_{ql}^{f}] \right.\right.$$

$$+ E[z_{pi}^{f} z_{pj}^{f\,*} z_{qk}^{b\,*} z_{ql}^{b}] + E[z_{pi}^{b} z_{pj}^{b\,*} z_{qk}^{f\,*} z_{ql}^{f}]$$

$$\left.\left. + E[z_{pi}^{b} z_{pj}^{b\,*} z_{qk}^{b\,*} z_{ql}^{b}]\right\}\right] - \tilde{\lambda}_i\tilde{\lambda}_k\delta_{ij}\delta_{kl}\,.$$

Using the results[2] for fourth order moments of jointly circular

---

(2) Let z_1, z_2, z_3, z_4 be jointly circular Gaussian random variables with zero mean. Then [9]

$$E[z_1 z_2^* z_3^* z_4] = E[z_1 z_2^*]E[z_4 z_3^*] + E[z_1 z_3^*]E[z_4 z_2^*].$$

Gaussian random variables and after some algebraic manipulations, we have

$$E[u_{ij}u_{kl}^*] =$$

$$= \frac{1}{N}\sum_{n=1}^{N}\frac{1}{4L^2}\sum_{p=1}^{L}\sum_{q=1}^{L}\left[E\,(z_{pi}^f z_{qk}^{f\,*})E\,(z_{ql}^f z_{pj}^{f\,*}) + E\,(z_{pi}^b z_{qk}^{b\,*})E\,(z_{ql}^b z_{pj}^{b\,*})\right.$$

$$+ E\,(z_{pi}^f z_{ql}^b)E\,(z_{pj}^f z_{qk}^b)^* + E\,(z_{pi}^b z_{ql}^f)E\,(z_{pj}^b z_{qk}^f)^*$$

$$\left. + \left\{E\,(z_{pi}^f z_{pj}^{f\,*}) + E\,(z_{pi}^b z_{pj}^{b\,*})\right\}\left\{E\,(z_{ql}^f z_{qk}^{f\,*}) + E\,(z_{ql}^b z_{qk}^{b\,*})\right\}\right] - \tilde{\lambda}_i\tilde{\lambda}_k\delta_{ij}\delta_{kl}$$

$$= \frac{1}{4L^2}\sum_{p=1}^{L}\sum_{q=1}^{L}\left(\tilde{\beta}_i^\dagger\mathbf{R}_{pq}^f\tilde{\beta}_k\,\tilde{\beta}_l^\dagger\mathbf{R}_{qp}^f\tilde{\beta}_j + \tilde{\beta}_i^\dagger\mathbf{R}_{pq}^b\tilde{\beta}_k\,\tilde{\beta}_l^\dagger\mathbf{R}_{qp}^b\tilde{\beta}_j\right.$$

$$\left. + \tilde{\beta}_i^\dagger\mathbf{R}_{pq_o}^f\tilde{\gamma}_l\,\tilde{\gamma}_j^\dagger\mathbf{R}_{p_o q}^b\tilde{\beta}_k + \tilde{\beta}_i^\dagger\mathbf{R}_{pq_o}^b\tilde{\gamma}_l\,\tilde{\gamma}_j^\dagger\mathbf{R}_{p_o q}^f\tilde{\beta}_k\right) \triangleq \tilde{\Gamma}_{iklj}\,. \tag{3.39}$$

Here by definition $p_o = M_o - M - p + 2$, $q_o = M_o - M - q + 2$ and $\tilde{\gamma}_j$ is the inverted $\tilde{\beta}_j^*$ vector with $\tilde{\gamma}_{j,m} = \tilde{\beta}_{j,M-m+1}^*$. For $L = 1$, in an uncorrelated as well as a two equipowered coherent source scene, it is easy to show that $\tilde{\beta}_i^\dagger\tilde{\mathbf{R}}\tilde{\beta}_j = \tilde{\gamma}_i^\dagger\tilde{\mathbf{R}}\tilde{\gamma}_j$ for all i,j. Using results in Appendix 3.B, it then follows that $\tilde{\gamma}_i = \tilde{\beta}_i\,e^{j\phi_i}$, $i = 1, 2, \cdots, K$ and $[\tilde{\gamma}_{K+1}, \cdots, \tilde{\gamma}_M] = [\tilde{\beta}_{K+1}, \cdots, \tilde{\beta}_M]\mathbf{V}$, where $\mathbf{V}$ is an $(M-K)\times(M-K)$ unitary matrix. Further,

$$\mathbf{R}_{pq}^f \triangleq E\,[\mathbf{x}_p^f(\mathbf{x}_q^f)^\dagger] \tag{3.40}$$

$$\mathbf{R}_{pq}^b \triangleq E\,[\mathbf{x}_p^b(\mathbf{x}_q^b)^\dagger]\,. \tag{3.41}$$

In obtaining (3.39), we have also made use of (3.38) and the fact that for circular Gaussian data

$$E\,[z_{pi}^f z_{qj}^{b\,*}] = \tilde{\beta}_i^\dagger E\,[\mathbf{x}_p^f(\mathbf{x}_q^b)^\dagger]\tilde{\beta}_j = \tilde{\beta}_i^\dagger E\,[\mathbf{x}_p^f(\mathbf{x}_{q_o}^f)^T]\gamma_j = 0\,,$$

$$p,q = 1, 2, \cdots, L\,. \tag{3.42}$$

The forward-only smoothing scheme now follows as a subclass of this general analysis. In that case the estimate of the smoothed covariance matrix is given by

$$\mathbf{S}_f = \frac{1}{L} \sum_{l=1}^{L} \left(\frac{1}{N} \sum_{n=1}^{N} \mathbf{x}_l^f(n)(\mathbf{x}_l^f(n))^\dagger \right) = \mathbf{E}_f \mathbf{L}_f \mathbf{E}_f^\dagger .$$

Then, as in (3.37) with $z_{li}^f(n) = \boldsymbol{\beta}_{fi}^\dagger \mathbf{x}_l^f(n)$ where $E[\mathbf{S}_f] = \mathbf{R}^f = \mathbf{B}_f \mathbf{\Lambda}_f \mathbf{B}_f^\dagger = \sum_{i=1}^{M} \lambda_{fi} \boldsymbol{\beta}_{fi} \boldsymbol{\beta}_{fi}^\dagger$, we have

$$u_{ij} = \frac{1}{\sqrt{N}} \sum_{n=1}^{N} \left(\frac{1}{L} \sum_{l=1}^{L} z_{li}^f z_{lj}^{f\,*} - \lambda_{fi} \delta_{ij} \right).$$

Thus, $E[u_{ij}] = 0$ and

$$E[u_{ij} u_{kl}^*] = \frac{1}{N} \sum_{n=1}^{N} \left[\frac{1}{L^2} \sum_{p=1}^{L} \sum_{q=1}^{L} E(z_{pi}^f z_{pj}^{f\,*} z_{qk}^{f\,*} z_{ql}^f) - \lambda_{fi} \lambda_{fk} \delta_{ij} \delta_{kl} \right]$$

$$= \frac{1}{L^2} \sum_{p=1}^{L} \sum_{q=1}^{L} \boldsymbol{\beta}_{fi}^\dagger \mathbf{R}_{pq}^f \boldsymbol{\beta}_{fk} \boldsymbol{\beta}_{fl}^\dagger \mathbf{R}_{qp}^f \boldsymbol{\beta}_{fj} \triangleq \Gamma_{iklj}^f = \Gamma_{ljik}^f \qquad (3.43)$$

where we have again made use of the circular Gaussian property of the data vectors.

Two important special cases of considerable practical interest are the conventional (unsmoothed) MUSIC scheme and the f/b smoothed scheme with $L = 1$ (see section 2.3.3). Of these, the former one corresponds to the forward-only scheme with $L = 1$, and in that case from (3.43)

$$\Gamma_{iklj} = \boldsymbol{\beta}_i^\dagger \mathbf{R} \boldsymbol{\beta}_k \boldsymbol{\beta}_l^\dagger \mathbf{R} \boldsymbol{\beta}_j = \lambda_i \lambda_j \delta_{ik} \delta_{jl} . \qquad (3.44)$$

In the latter case, using (3.40) and (3.41) together with the identity $\tilde{\boldsymbol{\beta}}_i^\dagger \mathbf{a}(\omega_k) = 0$ for $i = K+1, K+2, \cdots, M$ and $k = 1, 2, \cdots, K$ simplifies (3.39) into

$$\tilde{\Gamma}_{iklj} = \frac{1}{2} \tilde{\lambda}_i \tilde{\lambda}_j (\delta_{ik} \delta_{jl} + \tilde{\boldsymbol{\beta}}_i^\dagger \tilde{\boldsymbol{\gamma}}_l \tilde{\boldsymbol{\gamma}}_j^\dagger \tilde{\boldsymbol{\beta}}_k)$$

in an uncorrelated scene. Similarly, in a two equipowered coherent source scene, we also have

$$\tilde{\Gamma}_{iklj} = \begin{cases} \tilde{\lambda}_i \tilde{\lambda}_j \delta_{ik} \delta_{jl} - \dfrac{1}{4} \Big[\tilde{\beta}_i^{\dagger} \mathbf{R}_{11}^{f} \tilde{\beta}_k \tilde{\beta}_l^{\dagger} \mathbf{R}_{11}^{b} \tilde{\beta}_j + \tilde{\beta}_i^{\dagger} \mathbf{R}_{11}^{b} \tilde{\beta}_k \tilde{\beta}_l^{\dagger} \mathbf{R}_{11}^{f} \tilde{\beta}_j \\[2mm] \qquad - \tilde{\beta}_i^{\dagger} \mathbf{R}_{11}^{f} \tilde{\gamma}_l \tilde{\gamma}_j^{\dagger} \mathbf{R}_{11}^{b} \tilde{\beta}_k - \tilde{\beta}_i^{\dagger} \mathbf{R}_{11}^{b} \tilde{\gamma}_l \tilde{\gamma}_j^{\dagger} \mathbf{R}_{11}^{f} \tilde{\beta}_k \Big], \quad i,j,k,l \leq 2 \\[4mm] \dfrac{1}{2} \tilde{\lambda}_i \tilde{\lambda}_j (\delta_{ik} \delta_{jl} + \tilde{\beta}_i^{\dagger} \tilde{\gamma}_l \, \tilde{\gamma}_j^{\dagger} \tilde{\beta}_k), \qquad\qquad\qquad \text{otherwise}. \end{cases} \tag{3.45}$$

Proceeding as in (3.39), we have for the general case

$$E[u_{ij} u_{kl}] = \tilde{\Gamma}_{ilkj} . \tag{3.46}$$

Using (3.33) together with (3.34) and (3.31) we have

$$\mathbf{T} = \tilde{\mathbf{A}} + \frac{1}{\sqrt{N}} \mathbf{U} = \mathbf{Y} \tilde{\mathbf{L}} \mathbf{Y}^{\dagger} = \mathbf{Y} (\tilde{\mathbf{A}} + \frac{1}{\sqrt{N}} \mathbf{F}) \mathbf{Y}^{\dagger}$$

which gives the identity

$$\tilde{\mathbf{A}} + \frac{1}{\sqrt{N}} \mathbf{U} = \mathbf{Y} (\tilde{\mathbf{A}} + \frac{1}{\sqrt{N}} \mathbf{F}) \mathbf{Y}^{\dagger} . \tag{3.47}$$

To derive the asymptotic properties of the sample estimates corresponding to the distinct eigenvalues $\tilde{\lambda}_1, \tilde{\lambda}_2, \cdots, \tilde{\lambda}_K$ of $\tilde{\mathbf{R}}$, following [4, 7] we partition the matrices $\tilde{\mathbf{A}}, \mathbf{U}, \mathbf{F}$ and $\mathbf{Y}$ as follows

$$\tilde{\mathbf{A}} = \begin{bmatrix} \tilde{\mathbf{A}}_1 & \mathbf{O} \\ & \\ \mathbf{O} & \sigma^2 \mathbf{I}_{M-K} \end{bmatrix} , \quad \mathbf{U} = \begin{bmatrix} \mathbf{U}_{11} & \mathbf{U}_{12} \\ & \\ \mathbf{U}_{21} & \mathbf{U}_{22} \end{bmatrix}$$

$$\mathbf{F} = \begin{bmatrix} \mathbf{F}_1 & \mathbf{O} \\ \mathbf{O} & \mathbf{F}_2 \end{bmatrix}, \quad \mathbf{Y} = \begin{bmatrix} \mathbf{Y}_{11} & \mathbf{Y}_{12} \\ \mathbf{Y}_{21} & \mathbf{Y}_{22} \end{bmatrix}. \tag{3.48}$$

Here $\tilde{\mathbf{A}}_1$, $\mathbf{U}_{11}$, $\mathbf{F}_1$ and $\mathbf{Y}_{11}$ are of sizes $K{\times}K$, etc. With (3.48) in (3.47) and after some algebraic manipulations and retaining only those terms of order less than or equal to $1/\sqrt{N}$, we have

$$\begin{bmatrix} \tilde{\mathbf{A}}_1 & \mathbf{O} \\ \mathbf{O} & \sigma^2 \mathbf{I}_{M\text{-}K} \end{bmatrix} + \frac{1}{\sqrt{N}} \begin{bmatrix} \mathbf{U}_{11} & \mathbf{U}_{12} \\ \mathbf{U}_{21} & \mathbf{U}_{22} \end{bmatrix} = \begin{bmatrix} \tilde{\mathbf{A}}_1 & \mathbf{O} \\ \mathbf{O} & \sigma^2 \mathbf{Y}_{22}\mathbf{Y}_{22}^{\dagger} \end{bmatrix}$$

$$+ \frac{1}{\sqrt{N}}\left(\begin{bmatrix} \tilde{\mathbf{A}}_1 \mathbf{W}_{11}^{\dagger} & \tilde{\mathbf{A}}_1 \mathbf{W}_{21}^{\dagger} \\ \sigma^2 \mathbf{Y}_{22}\mathbf{W}_{12}^{\dagger} & \mathbf{O} \end{bmatrix} + \begin{bmatrix} \mathbf{F}_1 & \mathbf{O} \\ \mathbf{O} & \mathbf{Y}_{22}\mathbf{F}_2\mathbf{Y}_{22}^{\dagger} \end{bmatrix}\right.$$

$$\left.+ \begin{bmatrix} \mathbf{W}_{11}\tilde{\mathbf{A}}_1 & \sigma^2\mathbf{W}_{12}\mathbf{Y}_{22}^{\dagger} \\ \mathbf{W}_{21}\tilde{\mathbf{A}}_1 & \mathbf{O} \end{bmatrix}\right) + o\left(\frac{1}{N}\right) \tag{3.49}$$

where

$$\mathbf{W}_{11} = \sqrt{N}\,(\mathbf{Y}_{11} - \mathbf{I}_K) \tag{3.50}$$

$$\mathbf{W}_{12} = \sqrt{N}\,\mathbf{Y}_{12}, \quad \mathbf{W}_{21} = \sqrt{N}\,\mathbf{Y}_{21} \tag{3.51}$$

and define the column vectors $\mathbf{w}_1, \mathbf{w}_2, \cdots, \mathbf{w}_K$ from

$$\begin{bmatrix} \mathbf{W}_{11} \\ \mathbf{W}_{21} \end{bmatrix} = [\, \mathbf{w}_1, \mathbf{w}_2, \cdots, \mathbf{w}_K \,] \triangleq \mathbf{W}. \tag{3.52}$$

Similarly,

$$\mathbf{Y}\mathbf{Y}^\dagger = \mathbf{I}_M$$

$$= \begin{bmatrix} \mathbf{I}_K & \mathbf{O} \\ \\ \mathbf{O} & \mathbf{Y}_{22}\mathbf{Y}_{22}^\dagger \end{bmatrix} + \frac{1}{\sqrt{N}} \left(\begin{bmatrix} \mathbf{W}_{11} & \mathbf{W}_{12}\mathbf{Y}_{22}^\dagger \\ \\ \mathbf{W}_{21} & \mathbf{O} \end{bmatrix} + \begin{bmatrix} \mathbf{W}_{11}^\dagger & \mathbf{W}_{21}^\dagger \\ \\ \mathbf{Y}_{22}\mathbf{W}_{12}^\dagger & \mathbf{O} \end{bmatrix} \right)$$

$$+ \, o\,(1/N). \tag{3.53}$$

Thus, asymptotically for sufficiently large N, from (3.53) and (3.49) we have

$$\mathbf{O} = \mathbf{W}_{11} + \mathbf{W}_{11}^\dagger \tag{3.54}$$

$$\mathbf{W}_{21} + \mathbf{Y}_{22}\mathbf{W}_{12}^\dagger = \mathbf{O} \tag{3.55}$$

$$\mathbf{U}_{11} = \mathbf{W}_{11}\tilde{\mathbf{\Lambda}}_1 + \mathbf{F}_1 + \tilde{\mathbf{\Lambda}}_1 \mathbf{W}_{11}^\dagger \tag{3.56}$$

and

$$\mathbf{U}_{21} = \mathbf{W}_{21}\tilde{\mathbf{\Lambda}}_1 + \sigma^2 \mathbf{Y}_{22}\mathbf{W}_{12}^\dagger = \mathbf{W}_{21}\tilde{\mathbf{\Lambda}}_1 - \sigma^2 \mathbf{W}_{21}. \tag{3.57}$$

Since $y_{ii} \geq 0$, this together with (3.50) implies w_{ii} is real and hence from (3.54)

$$w_{ii} = 0, \quad i = 1, 2, \cdots, K$$

and

$$w_{ij} = -w_{ji}^*, \quad i, j = 1, 2, \cdots, K, \quad i \neq j$$

which when substituted into (3.56) $-$ (3.57) gives

$$f_{ii} = u_{ii}, \quad i = 1, 2, \cdots, K \tag{3.58}$$

$$
w_{ij} = \begin{cases} \dfrac{u_{ij}}{(\tilde{\lambda}_j - \tilde{\lambda}_i)} & i,j = 1, 2, \cdots, K, \quad i \neq j \\[3ex] \dfrac{u_{ij}}{\tilde{\lambda}_j - \sigma^2} & i = K+1, K+2, \cdots, M, \quad j = 1, 2, \cdots, K. \end{cases} \tag{3.59}
$$

From (3.32) and (3.27) we also have

$$
\mathbf{G} = \sqrt{N}\,(\mathbf{C} - \tilde{\mathbf{B}}) = \sqrt{N}\,\tilde{\mathbf{B}}\,(\mathbf{Y} - \mathbf{I}) = \sqrt{N}\,\tilde{\mathbf{B}} \begin{bmatrix} \mathbf{Y}_{11} - \mathbf{I}_K & \mathbf{Y}_{12} \\[2ex] \mathbf{Y}_{21} & \mathbf{Y}_{22} - \mathbf{I}_{M\text{-}K} \end{bmatrix}
$$

which together with (3.50) $-$ (3.52) gives

$$
[\mathbf{g}_1, \mathbf{g}_2, \cdots, \mathbf{g}_K] = \tilde{\mathbf{B}}\,[\mathbf{w}_1, \mathbf{w}_2, \cdots, \mathbf{w}_K]
$$

or

$$
\mathbf{g}_i = \sqrt{N}\,(\mathbf{c}_i - \tilde{\boldsymbol{\beta}}_i) = \tilde{\mathbf{B}}\,\mathbf{w}_i, \quad i = 1, 2, \cdots, K.
$$

Using (3.31) and (3.58), for $i = 1, 2, \cdots, K$, this yields

$$
\tilde{l}_i = \tilde{\lambda}_i + (1/\sqrt{N})f_{ii} = \tilde{\lambda}_i + (1/\sqrt{N})u_{ii}, \tag{3.60}
$$

and

$$
\mathbf{c}_i = \tilde{\boldsymbol{\beta}}_i + (1/\sqrt{N})\tilde{\mathbf{B}}\mathbf{w}_i = \tilde{\boldsymbol{\beta}}_i + (1/\sqrt{N}) \sum_{\substack{j=1 \\ j \neq i}}^{M} w_{ji}\tilde{\boldsymbol{\beta}}_j. \tag{3.61}
$$

Thus, the estimators $\tilde{l}_i$ and $\mathbf{c}_i$, $i = 1, 2, \cdots, K$, corresponding to the largest K eigenvalues of $\tilde{\mathbf{R}}$, are asymptotically multivariate Gaussian random variables/vectors with mean values $\tilde{\lambda}_i$ and $\tilde{\boldsymbol{\beta}}_i$, $i = 1, 2, \cdots, K$, respectively. Further,

$$
Cov(\tilde{l}_i, \tilde{l}_j) = \frac{1}{N} E[u_{ii}u_{jj}] = \frac{1}{N}\tilde{\Gamma}_{ijji}, \quad i,j = 1, 2, \cdots, K \tag{3.62}
$$

and

$$Cov(\mathbf{c}_i, \mathbf{c}_j) = \frac{1}{N} \sum_{\substack{k=1 \\ k \neq i}}^{M} \sum_{\substack{l=1 \\ l \neq j}}^{M} E[w_{ki} w_{lj}^*] \tilde{\beta}_k \tilde{\beta}_l^\dagger$$

$$= \frac{1}{N} \sum_{\substack{k=1 \\ k \neq i}}^{M} \sum_{\substack{l=1 \\ l \neq j}}^{M} \frac{\tilde{\Gamma}_{klji}}{(\tilde{\lambda}_i - \tilde{\lambda}_k)(\tilde{\lambda}_j - \tilde{\lambda}_l)} \tilde{\beta}_k \tilde{\beta}_l^\dagger. \tag{3.63}$$

Notice that $\mathbf{c}_i$ in (3.61) are not normalized vectors, and it may be emphasized that in the case of eigenvectors, the above asymptotic joint Gaussian property only holds good for these specific sets of unnormalized sample estimators. However, from (3.26) since the eigenvectors $\mathbf{c}_i$, $i = 1, 2, \cdots, K$, appearing in (3.29) are normalized ones, to make use of the explicit forms given by (3.61) there, we proceed to normalize these vectors. Starting from (3.61), we have

$$\| \mathbf{c}_i \|^2 = \mathbf{c}_i^\dagger \mathbf{c}_i = 1 + \frac{1}{N} \sum_{\substack{j=1 \\ j \neq i}}^{M} |w_{ji}|^2 > 1 \tag{3.64}$$

and, hence, the corresponding normalized eigenvectors $\hat{\mathbf{c}}_i$, $i = 1, 2, \cdots, K$ have the form

$$\hat{\mathbf{c}}_i \triangleq \| \mathbf{c}_i \|^{-1} \mathbf{c}_i = \left[1 + \frac{1}{N} \sum_{\substack{j=1 \\ j \neq i}}^{M} |w_{ji}|^2 \right]^{-1/2} \left[\tilde{\beta}_i + \frac{1}{\sqrt{N}} \sum_{\substack{j=1 \\ j \neq i}}^{M} w_{ji} \tilde{\beta}_j \right]$$

$$= \left[1 - \frac{1}{2N} \sum_{\substack{j=1 \\ j \neq i}}^{M} |w_{ji}|^2 \right] \tilde{\beta}_i + \frac{1}{\sqrt{N}} \sum_{\substack{j=1 \\ j \neq i}}^{M} w_{ji} \tilde{\beta}_j$$

$$- \frac{1}{2N\sqrt{N}} \sum_{\substack{k=1 \\ k \neq i}}^{M} \sum_{\substack{l=1 \\ l \neq i}}^{M} |w_{ki}|^2 w_{li} \tilde{\beta}_l + o(1/N^2). \tag{3.65}$$

Using (3.59) and (3.39) we have

$$E[\hat{c}_i] = \tilde{\beta}_i - \frac{1}{2N} \sum_{\substack{j=1 \\ j \neq i}}^{M} E[\,|w_{ji}|^2]\tilde{\beta}_i + o(1/N^2)$$

$$= \tilde{\beta}_i - \frac{1}{2N} \sum_{\substack{j=1 \\ j \neq i}}^{M} \frac{\tilde{\Gamma}_{iijj}}{(\tilde{\lambda}_i - \tilde{\lambda}_j)^2}\tilde{\beta}_i + o(1/N^2),$$

$$i = 1, 2, \cdots, K, \tag{3.66}$$

since from the asymptotic joint normal distribution of these zero mean random variables u_{ij} $(i \neq j)$, their odd order moments are zero. Thus, asymptotically, these normalized estimates for the eigenvectors $\tilde{\beta}_1, \tilde{\beta}_2, \cdots, \tilde{\beta}_K$, of $\tilde{R}$ are unbiased, and the exact bias expressions are given by (3.66). Further,

$$\hat{c}_i\hat{c}_j^\dagger = \left[1 - \frac{1}{2N}\left(\sum_{\substack{k=1 \\ k \neq i}}^{M} |w_{ki}|^2 + \sum_{\substack{k=1 \\ k \neq j}}^{M} |w_{kj}|^2\right)\right]\tilde{\beta}_i\tilde{\beta}_j^\dagger$$

$$+ \frac{1}{\sqrt{N}}\left(\sum_{\substack{k=1 \\ k \neq i}}^{M} w_{ki}\tilde{\beta}_k\tilde{\beta}_j^\dagger + \sum_{\substack{k=1 \\ k \neq j}}^{M} w_{kj}^*\tilde{\beta}_i\tilde{\beta}_k^\dagger\right) + \frac{1}{N} \sum_{\substack{k=1 \\ k \neq i}}^{M} \sum_{\substack{l=1 \\ l \neq j}}^{M} w_{ki}w_{lj}^*\tilde{\beta}_k\tilde{\beta}_l^\dagger$$

$$- \frac{1}{2N\sqrt{N}}\left(\sum_{\substack{k=1 \\ k \neq i}}^{M} \sum_{\substack{l=1 \\ l \neq i}}^{M} |w_{ki}|^2 w_{li}\tilde{\beta}_l\tilde{\beta}_j^\dagger + \sum_{\substack{k=1 \\ k \neq j}}^{M} \sum_{\substack{l=1 \\ l \neq j}}^{M} |w_{kj}|^2 w_{lj}^*\tilde{\beta}_i\tilde{\beta}_l^\dagger\right.$$

$$+ \sum_{\substack{k=1 \\ k \neq i}}^{M} \sum_{\substack{l=1 \\ l \neq j}}^{M} w_{ki}|w_{lj}|^2\tilde{\beta}_k\tilde{\beta}_j^\dagger + \left.\sum_{\substack{k=1 \\ k \neq i}}^{M} \sum_{\substack{l=1 \\ l \neq j}}^{M} |w_{ki}|^2 w_{lj}^*\tilde{\beta}_i\tilde{\beta}_l^\dagger\right)$$

$$+ o(1/N^2). \tag{3.67}$$

Again, neglecting terms of order $1/N^2$ and proceeding as above, this expression reduces to

$$E[\hat{c}_i\,\hat{c}_j^\dagger] = \tilde{\beta}_i\tilde{\beta}_j^\dagger - \frac{1}{2N}\left[\sum_{\substack{k=1\\k\neq i}}^{M} E[\,|w_{ki}|^2\,] + \sum_{\substack{k=1\\k\neq j}}^{M} E[\,|w_{kj}|^2\,]\right]\tilde{\beta}_i\tilde{\beta}_j^\dagger$$

$$+ \frac{1}{N}\sum_{\substack{k=1\\k\neq i}}^{M}\sum_{\substack{l=1\\l\neq j}}^{M} E[w_{ki}w_{lj}^*]\tilde{\beta}_k\tilde{\beta}_l^\dagger + o\,(1/N^2)$$

$$= \tilde{\beta}_i\tilde{\beta}_j^\dagger - \frac{1}{2N}\left[\sum_{\substack{k=1\\k\neq i}}^{M}\frac{\tilde{\Gamma}_{iikk}}{(\tilde{\lambda}_i - \tilde{\lambda}_k)^2} + \sum_{\substack{k=1\\k\neq j}}^{M}\frac{\tilde{\Gamma}_{jjkk}}{(\tilde{\lambda}_j - \tilde{\lambda}_k)^2}\right]\tilde{\beta}_i\tilde{\beta}_j^\dagger$$

$$+ \frac{1}{N}\sum_{\substack{k=1\\k\neq i}}^{M}\sum_{\substack{l=1\\l\neq j}}^{M}\frac{\tilde{\Gamma}_{klji}}{(\tilde{\lambda}_i - \tilde{\lambda}_k)(\tilde{\lambda}_j - \tilde{\lambda}_l)}\tilde{\beta}_k\tilde{\beta}_l^\dagger + o\,(1/N^2) \quad (3.68)$$

and in a similar manner, using (3.46)

$$E[\hat{c}_i\,\hat{c}_j^T] = \tilde{\beta}_i\tilde{\beta}_j^T - \frac{1}{2N}\left[\sum_{\substack{k=1\\k\neq i}}^{M}\frac{\tilde{\Gamma}_{iikk}}{(\tilde{\lambda}_i - \tilde{\lambda}_k)^2} + \sum_{\substack{k=1\\k\neq j}}^{M}\frac{\tilde{\Gamma}_{jjkk}}{(\tilde{\lambda}_j - \tilde{\lambda}_k)^2}\right]\tilde{\beta}_i\tilde{\beta}_j^T$$

$$+ \frac{1}{N}\sum_{\substack{k=1\\k\neq i}}^{M}\sum_{\substack{l=1\\l\neq j}}^{M}\frac{\tilde{\Gamma}_{kjli}}{(\tilde{\lambda}_i - \tilde{\lambda}_k)(\tilde{\lambda}_j - \tilde{\lambda}_l)}\tilde{\beta}_k\tilde{\beta}_l^T + o\,(1/N^2)\,. \quad (3.69)$$

An easy verification shows that $Cov\,(\hat{c}_i, \hat{c}_j)$ is once again given by (3.63), but nevertheless, (3.67) − (3.68) will turn out to be useful in computing the asymptotic bias and variance of the sample direction estimator $\hat{Q}(\omega)$ in (3.29).

These general expressions in (3.66) − (3.69) for first and second order moments of normalized eigenvector estimators can be used to evaluate their counterparts for the forward-only and the conventional MUSIC cases by substituting the proper Γ values derived in (3.43) and

(3.44), respectively. Thus, for the conventional MUSIC case, using (3.44) we have

$$E[\hat{c}_i] = \beta_i - \frac{1}{2N} \sum_{\substack{j=1 \\ j \neq i}}^{M} \frac{\lambda_i \lambda_j}{(\lambda_i - \lambda_j)^2} \beta_i + o(1/N^2) \qquad (3.70)$$

$$E[\hat{c}_i \hat{c}_j^\dagger] = \beta_i \beta_j^\dagger - \frac{1}{2N} \left[\sum_{\substack{k=1 \\ k \neq i}}^{M} \frac{\lambda_i \lambda_k}{(\lambda_i - \lambda_k)^2} + \sum_{\substack{k=1 \\ k \neq j}}^{M} \frac{\lambda_j \lambda_k}{(\lambda_j - \lambda_k)^2} \right] \beta_i \beta_j^\dagger$$

$$+ \frac{1}{N} \sum_{\substack{k=1 \\ k \neq i}}^{M} \frac{\lambda_i \lambda_k}{(\lambda_i - \lambda_k)^2} \beta_k \beta_k^\dagger \delta_{ij} + o(1/N^2) \qquad (3.71)$$

and

$$E[\hat{c}_i \hat{c}_j^T] = \beta_i \beta_j^T - \frac{1}{2N} \left[\sum_{\substack{k=1 \\ k \neq i}}^{M} \frac{\lambda_i \lambda_k}{(\lambda_i - \lambda_k)^2} + \sum_{\substack{k=1 \\ k \neq j}}^{M} \frac{\lambda_j \lambda_k}{(\lambda_j - \lambda_k)^2} \right] \beta_i \beta_j^T$$

$$- \frac{1}{N} \frac{\lambda_i \lambda_j}{(\lambda_i - \lambda_j)^2} \beta_j \beta_i^T (1 - \delta_{ij}) + o(1/N^2) ,$$

where $\lambda_i, \beta_i, i = 1, 2, \cdots, M$ are as defined in (2.60).

Once again for the f/b smoothed case using (3.29) − (3.30) and recalling that the eigenvector estimators appearing there are normalized ones, we have

$$\hat{y}_i(\omega) = |\hat{c}_i^\dagger a(\omega)|^2 = a^\dagger(\omega)\hat{c}_i \hat{c}_i^\dagger a(\omega) \qquad (3.72)$$

and from (3.68) and (3.29) we have

$$E[\hat{Q}(\omega)] = \tilde{Q}(\omega) + \frac{1}{N} \sum_{i=1}^{K} \left[\sum_{\substack{k=1 \\ k \neq i}}^{M} \frac{\tilde{\Gamma}_{iikk}}{(\tilde{\lambda}_i - \tilde{\lambda}_k)^2} \, |\, \tilde{\beta}_i^\dagger \mathbf{a}(\omega)\, |^2 \right.$$

$$\left. - \sum_{\substack{k=1 \\ k \neq i}}^{M} \sum_{\substack{l=1 \\ l \neq i}}^{M} \frac{\tilde{\Gamma}_{iikl}}{(\tilde{\lambda}_i - \tilde{\lambda}_k)(\tilde{\lambda}_i - \tilde{\lambda}_l)} \, \mathbf{a}^\dagger(\omega)\tilde{\beta}_k \tilde{\beta}_l^\dagger \mathbf{a}(\omega) \right] + o\,(1/N^2) \quad (3.73)$$

and this shows that $E[\hat{Q}(\omega)] \to \tilde{Q}(\omega)$ as $N \to \infty$. Similar bias expressions for the forward-only smoothing scheme can be obtained from (3.73) by replacing $\tilde{\Gamma}_{ijkl}$ with (3.43). In particular for the conventional (unsmoothed) MUSIC case with (3.44) in (3.73), after some simplifications, it reduces to

$$E[\hat{Q}(\omega)]$$

$$= Q(\omega) + \frac{1}{N} \sum_{i=1}^{K} \left[\sum_{\substack{k=1 \\ k \neq i}}^{M} \frac{\lambda_i \lambda_k}{(\lambda_i - \lambda_k)^2} \left(\,|\, \beta_i^\dagger \mathbf{a}(\omega)\,|^2 - \,|\, \beta_k^\dagger \mathbf{a}(\omega)\,|^2 \right) \right]$$

$$+ o\,(1/N^2)$$

$$= Q(\omega) + \frac{1}{N} \sum_{i=1}^{K} \frac{\lambda_i \sigma^2}{(\lambda_i - \sigma^2)^2} \left[(M - K)\, |\, \beta_i^\dagger \mathbf{a}(\omega)\,|^2 - Q(\omega) \right]$$

$$+ o\,(1/N^2) \qquad (3.74)$$

where $\lambda_i, \beta_i, i = 1, 2, \cdots, M$, are as defined in (2.60) and $Q(\omega)$ is given by (2.67). Also, from (3.29)

$$Var(\hat{Q}(\omega)) = E[\hat{Q}^2(\omega)] - \left(E[\hat{Q}(\omega)] \right)^2$$

$$= E[(1 - \sum_{i=1}^{K} \hat{y}_i)^2] - \left[E[1 - \sum_{i=1}^{K} \hat{y}_i] \right]^2$$

$$= \sum_{i=1}^{K} \sum_{j=1}^{K} \left(E[\hat{y}_i \hat{y}_j] - E[\hat{y}_i]E[\hat{y}_j] \right). \qquad (3.75)$$

Using (3.72) and (3.67), after a series of manipulations, we have

$$E[\hat{y}_i\hat{y}_j] - E[\hat{y}_i]E[\hat{y}_j] = \frac{1}{N}E[d_i(\omega)d_j(\omega)] + o\,(1/N^2)$$

where

$$d_i(\omega) = \sum_{\substack{k=1\\k\neq i}}^{M}\left(w_{ki}\mathbf{a}^\dagger(\omega)\tilde{\boldsymbol{\beta}}_k\tilde{\boldsymbol{\beta}}_i^\dagger\mathbf{a}(\omega) + w_{ki}^*\mathbf{a}^\dagger(\omega)\tilde{\boldsymbol{\beta}}_i\tilde{\boldsymbol{\beta}}_k^\dagger\mathbf{a}(\omega)\right)$$

which gives

$$E[\hat{y}_i\hat{y}_j] - E[\hat{y}_i]E[\hat{y}_j] = \frac{2}{N}\sum_{\substack{k=1\\k\neq i}}^{M}\sum_{\substack{l=1\\l\neq j}}^{M}$$

$$\frac{\mathrm{Re}\left[\left(\bar{\Gamma}_{klji}\mathbf{a}^\dagger(\omega)\tilde{\boldsymbol{\beta}}_j\tilde{\boldsymbol{\beta}}_l^\dagger\mathbf{a}(\omega) + \bar{\Gamma}_{kjli}\mathbf{a}^\dagger(\omega)\tilde{\boldsymbol{\beta}}_l\tilde{\boldsymbol{\beta}}_j^\dagger\mathbf{a}(\omega)\right)\mathbf{a}^\dagger(\omega)\tilde{\boldsymbol{\beta}}_k\tilde{\boldsymbol{\beta}}_i^\dagger\mathbf{a}(\omega)\right]}{(\tilde{\lambda}_i - \tilde{\lambda}_k)(\tilde{\lambda}_j - \tilde{\lambda}_l)}$$

$$+ o\,(1/N^2). \qquad (3.76)$$

Finally, with (3.76) in (3.75) we get

$$Var(\hat{Q}(\omega)) = \frac{2}{N}\sum_{i=1}^{K}\sum_{j=1}^{K}\sum_{\substack{k=1\\k\neq i}}^{M}\sum_{\substack{l=1\\l\neq j}}^{M}$$

$$\frac{\mathrm{Re}\left[\left(\bar{\Gamma}_{klji}\mathbf{a}^\dagger(\omega)\tilde{\boldsymbol{\beta}}_j\tilde{\boldsymbol{\beta}}_l^\dagger\mathbf{a}(\omega) + \bar{\Gamma}_{kjli}\mathbf{a}^\dagger(\omega)\tilde{\boldsymbol{\beta}}_l\tilde{\boldsymbol{\beta}}_j^\dagger\mathbf{a}(\omega)\right)\mathbf{a}^\dagger(\omega)\tilde{\boldsymbol{\beta}}_k\tilde{\boldsymbol{\beta}}_i^\dagger\mathbf{a}(\omega)\right]}{(\tilde{\lambda}_i - \tilde{\lambda}_k)(\tilde{\lambda}_j - \tilde{\lambda}_l)}$$

$$+ o\,(1/N^2) \qquad (3.77)$$

Thus,

$$Var(\hat{Q}(\omega)) \to 0 \quad \text{as} \quad N \to \infty.$$

and $\hat{Q}(\omega)$ is a consistent estimator in all cases. For the conventional MUSIC case, (3.77) reduces to

$$Var(\hat{Q}(\omega)) \triangleq \sigma^2(\omega)$$

$$= \frac{2}{N} \sum_{i=1}^{K} \left[\sum_{\substack{k=1 \\ k \neq i}}^{M} \frac{\lambda_i \lambda_k}{(\lambda_i - \lambda_k)^2} \mid \beta_i^\dagger a(\omega) \mid^2 \mid \beta_k^\dagger a(\omega) \mid^2 \right.$$

$$\left. - \sum_{\substack{j=1 \\ j \neq i}}^{K} \frac{\lambda_i \lambda_j}{(\lambda_i - \lambda_j)^2} \mid \beta_i^\dagger a(\omega) \mid^2 \mid \beta_j^\dagger a(\omega) \mid^2 \right] + o(1/N^2)$$

$$= \frac{2}{N} \sum_{i=1}^{K} \sum_{k=K+1}^{M} \frac{\lambda_i \lambda_k}{(\lambda_i - \lambda_k)^2} \mid \beta_i^\dagger a(\omega) \mid^2 \mid \beta_k^\dagger a(\omega) \mid^2 + o(1/N^2)$$

$$= \frac{2}{N} Q(\omega) \sum_{i=1}^{K} \frac{\lambda_i \sigma^2}{(\lambda_i - \sigma^2)^2} \mid \beta_i^\dagger a(\omega) \mid^2 + o(1/N^2). \tag{3.78}$$

Since along the actual arrival angles, $Q(\omega_k) = 0$, $k = 1, 2, \cdots, K$, (3.78) allows us to conclude that within the above approximation,

$$\sigma^2(\omega_k) = 0, \quad k = 1, 2, \cdots, K, \tag{3.79}$$

i.e., in all multiple target situations, where the conventional MUSIC scheme is applicable, the variance of the estimator in (3.23) along the true arrival angles is zero within a first order approximation.

The general expressions for bias and variance in (3.73) and (3.77) can be used to determine the required sample size for a certain performance level or to arrive at useful resolution criteria for the forward-only or the f/b smoothing schemes. Though the general cases are often intractable, a complete analysis is possible for the f/b scheme with $L = 1$, which of course can decorrelate and resolve two coherent sources. As shown in the next section, this case leads to some interesting results, including the resolution threshold for two completely coherent equipowered sources in white noise.

3.3.2 Two Source Case – Uncorrelated and Coherent Scene

If the two sources present in the scene are coherent, then the array output data together with its complex conjugated backward version $x^b(n)$; $n = 1, 2, \cdots, N$, (f/b smoothing scheme with $L = 1$) can be used to decorrelate the incoming signals, thereby making it possible to estimate their arrival angles. In that case for two equipowered sources, the bias and variance of the sample estimator for the arrival angles can be computed by using (3.45) in (3.73) and (3.77), respectively. After a series of algebraic manipulations, for mean value of the estimator we have

$$E[\hat{Q}(\omega)]$$

$$= \tilde{Q}(\omega) + \frac{1}{2N} \sum_{i=1}^{2} \frac{\tilde{\lambda}_i \sigma^2}{(\tilde{\lambda}_i - \sigma^2)^2} \left[(M - 2) \, | \, \tilde{\beta}_i^\dagger \mathbf{a}(\omega) \, |^2 - \tilde{Q}(\omega) \right]$$

$$+ \, o\,(1/N^2). \tag{3.80}$$

Let $\tilde{\eta}(\omega)$ and $\eta(\omega)$ denote the bias in the f/b smoothing scheme and the conventional MUSIC scheme. Then from (3.80)

$$\tilde{\eta}(\omega) \triangleq E[\hat{Q}(\omega)] - \tilde{Q}(\omega)$$

$$= \frac{1}{2N} \sum_{i=1}^{2} \frac{\tilde{\lambda}_i \sigma^2}{(\tilde{\lambda}_i - \sigma^2)^2} \left[(M - 2) \, | \, \tilde{\beta}_i^\dagger \mathbf{a}(\omega) \, |^2 - \tilde{Q}(\omega) \right]$$

$$+ \, o\,(1/N^2) \tag{3.81}$$

and from (3.74) with $K = 2$ we have

$$\eta(\omega) = \frac{1}{N} \sum_{i=1}^{2} \frac{\lambda_i \sigma^2}{(\lambda_i - \sigma^2)^2} \left[(M - 2) \, | \, \beta_i^\dagger \mathbf{a}(\omega) \, |^2 - Q(\omega) \right]$$

$$+ \, o\,(1/N^2). \tag{3.82}$$

Notice that $\eta(\omega_k) > 0$, $\tilde{\eta}(\omega_k) > 0$, $k = 1, 2, \cdots, K$. To evaluate variance, let q_{ijkl} denote a typical term in (3.77). Then

$$\tilde{\sigma}^2(\omega) = \frac{2}{N}(q_{1122} + q_{1221} + q_{2112} + q_{2211}) + \frac{2}{N}\sum_{i=1}^{2}\sum_{j=1}^{2}\sum_{k=3}^{M}\sum_{l=3}^{M} q_{ijkl} \cdot$$

Using (3.45) it is easy to show that the terms within the first parentheses add up to zero, and by repeated use of (3.45) over the remaining terms we get

$$\tilde{\sigma}^2(\omega) = \frac{1}{N}\sum_{i=1}^{2}\frac{\tilde{\lambda}_i\sigma^2}{(\tilde{\lambda}_i - \sigma^2)^2}\Bigg[\; |\,\tilde{\beta}_i^\dagger \mathbf{a}(\omega)\,|^2\, \tilde{Q}(\omega)$$

$$+ \sum_{k=3}^{M} \mathrm{Re}\Big[\tilde{\beta}_k^\dagger\, \mathbf{a}(\omega)\, \tilde{\gamma}_k^\dagger\, \mathbf{a}(\omega)\, \mathbf{a}^\dagger(\omega)\, \tilde{\beta}_i\, \mathbf{a}^\dagger(\omega)\, \tilde{\gamma}_i\Big]\Bigg] + o(1/N^2)$$

$$= \frac{2\tilde{Q}(\omega)}{N}\sum_{i=1}^{2}\frac{\tilde{\lambda}_i\sigma^2}{(\tilde{\lambda}_i - \sigma^2)^2}\,|\,\tilde{\beta}_i^\dagger \mathbf{a}(\omega)\,|^2 + o(1/N^2). \qquad (3.83)$$

Again $\tilde{\sigma}^2(\omega_1) = \tilde{\sigma}^2(\omega_2) = 0$ in this case also. Notice that bias expressions in (3.81) and (3.82) for the f/b smoothing scheme with $L = 1$ and the conventional MUSIC scheme with $K = 2$ are functionally identical except for a multiplication factor of two in the conventional case. Moreover, these results suggest that in a correlated two source case, the f/b scheme will perform superior to the conventional one. This can be easily illustrated in an uncorrelated scene where equality of the array output covariance matrices in the smoothed and conventional cases implies $\lambda_i = \tilde{\lambda}_i$, $\beta_i = \tilde{\beta}_i$, $i = 1, 2, \cdots, M$ and $\tilde{\Gamma}_{iklj} = (\lambda_i\lambda_j/2)(\delta_{ik}\delta_{jl} + \beta_i^\dagger\gamma_l\,\gamma_j^\dagger\beta_k)$. Substituting these in (3.73) and (3.77), it easily follows that $\tilde{\eta}(\omega) = \eta(\omega)/2$ and $\tilde{\sigma}^2(\omega) = \sigma^2(\omega)$ and consequently, the f/b scheme is uniformly superior to the conventional one. However, in a correlated scene, although the effective correlation coefficient reduces in magnitude after smoothing (problem 6, chapter 2), it is difficult to exhibit such uniform superior behavior explicitly. Nevertheless, certain clarifications are possible. To see this, consider two cases, the first one consisting of two correlated sources with correlation coefficient ρ_o and the second one consisting of two coherent sources. In both cases, the sources are of equal power and have the same arrival angles θ_1 and θ_2. The correlated case can be

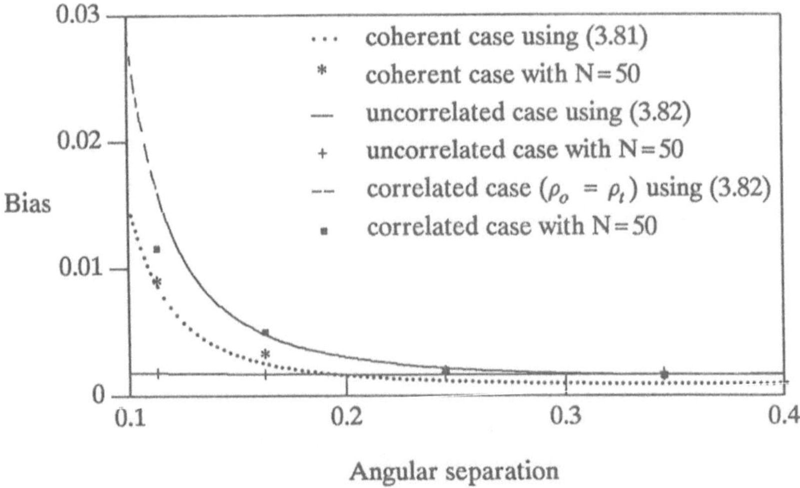

Fig. 3.1 Bias at one of the arrival angles *vs.* angular separation for two equipowered sources in uncorrelated, correlated and coherent scenes. A ten-element array is used to collect signal in all these cases. Input SNR is taken to be 10dB.

resolved using the conventional MUSIC scheme, and the coherent scene can be decorrelated and resolved using the f/b smoothing scheme with $L = 1$. In the later case from (3.C.3) and in particular from (3.C.14), smoothing results in an effective correlation coefficient $\rho_t = e^{j(M-1)\omega_d} \cos(M-1)\omega_d$ with $\omega_d = \pi \ (\cos\theta_1 - \cos\theta_2)/2$, between the sources. In the event that the temporal correlation ρ_o in the conventional case is equal to the above ρ_t, then $\mathbf{R} = \tilde{\mathbf{R}}$, $\lambda_i = \tilde{\lambda}_i$; $\beta_i = \tilde{\beta}_i$, $i = 1, 2$; and from (3.78), (3.81) − (3.83) the f/b scheme is uniformly superior to the conventional one in terms of bias. This conclusion is also supported by simulation results presented in Fig. 3.1 with details as indicated there.

As one would expect, for closely spaced sources the performance of the conventional scheme in an uncorrelated source scene is superior to that of the f/b scheme in a coherent scene. This is to be expected because for small values of angular separation ($\Delta^2 < 1$) from (3.C.39) and (3.C.40), we have $\eta(\omega_i) < \tilde{\eta}(\omega_i), i = 1, 2$. The deviation of $\eta(\omega_i)$ and $\tilde{\eta}(\omega_i)$ from zero − their nominal value − suggests the loss in resolution for the respective estimators. Since the estimators have zero

Table 3.1 Resolution threshold and probability of resolution *vs.* angular separation for two equipowered sources in uncorrelated and coherent scenes (number of sensors = 7, number of snapshots = 100, number of simulations = 100).

Angles of arrival			Uncorrelated		Coherent	
θ_1	θ_2	$2\omega_d$	SNR(dB)	Prob.	SNR(dB)	Prob.
19.00	25.00	0.1232	17	0.16	26	0.17
			18	0.17	27	0.39
			19	0.22	28	0.44
			20	0.47	29	0.58
			21	0.51		
			22	0.71		
34.00	40.00	0.1978	11	0.26	15	0.27
			12	0.39	16	0.35
			13	0.60	17	0.61
			14	0.65	18	0.76
113.00	118.00	0.2474	6	0.20	9	0.09
			7	0.30	10	0.36
			8	0.41	11	0.44
			9	0.52	12	0.57
			10	0.74		
60.00	66.00	0.2930	4	0.23	5	0.22
			5	0.37	6	0.36
			6	0.47	7	0.35
			7	0.66	8	0.62
127.00	135.00	0.3308	1	0.18	1	0.11
			2	0.22	2	0.20
			3	0.33	3	0.26
			4	0.53	4	0.47
			5	0.66	5	0.60

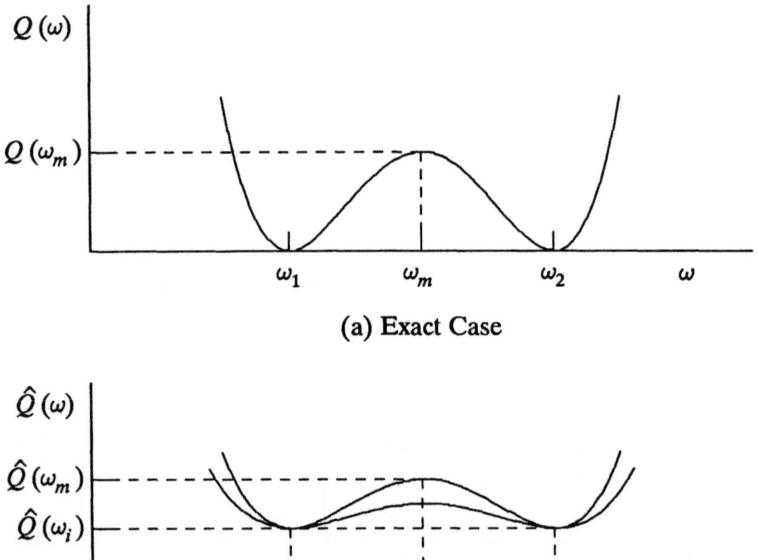

(a) Exact Case

(b) Estimated Case

Fig. 3.2 Resolution threshold analysis for MUSIC-type schemes.

variance along the two arrival angles in both cases, for a fixed number of samples a threshold in terms of SNR exists below which the two nulls corresponding to the true arrival angles are no longer separately identifiable. This has led to the definition of the resolution threshold [10] for two closely spaced sources as that value of SNR at which (see Fig. 3.2)

$$E[\hat{Q}(\omega_1)] = E[\hat{Q}(\omega_2)] = E[\hat{Q}((\omega_1 + \omega_2)/2)] , \qquad (3.84)$$

whenever $Var(\hat{Q}(\omega_1)) = Var(\hat{Q}(\omega_2)) = 0$. In the case of two equi-powered uncorrelated sources equating (3.C.39) and (3.C.41), Kaveh et $al.$ found the resolution threshold to be [10]

$$\xi_T \simeq \frac{1}{N} \left[\frac{20(M-2)}{\Delta^4} \left(1 + \left(1 + \frac{N}{5(M-2)} \Delta^2 \right)^{1/2} \right) \right] . \qquad (3.85)$$

Similarly the corresponding threshold $\tilde{\xi}_{\xi_T}$ in the coherent case can be found by equating (3.C.40) and (3.C.42). In that case [11]

$$\tilde{\xi}_T \simeq \frac{1}{N}\left[\frac{20(M-2)}{\Delta^4}\left(\frac{1}{3\Delta^2} - \frac{1}{20} - \frac{\Delta^2}{16}\right)\right.$$

$$\left.\left[1 + \left[1 + \frac{N(M-2)}{5M(M-4)}\Delta^2\right]^{1/2}\right]\right]$$

$$= \left(\frac{1}{3\Delta^2} - \frac{1}{20} - \frac{\Delta^2}{16}\right)\xi_T \simeq \left(\frac{1}{M\omega_d}\right)^2 \xi_T. \qquad (3.86)$$

Though $\tilde{\xi}_{\xi_T}$ and ξ_T possess similar features, for small arrays the resolution threshold in the coherent case can be substantially larger than that in the uncorrelated case. This asymptotic analysis is also found to be in agreement with results obtained from Monte Carlo simulations. A typical case study is reported in Table 3.1. When the equality in (3.84) is true, the probability of resolution was found to range from 0.33 to 0.5 in both cases there. This in turn implies that the above analysis should give an approximate threshold in terms of ξ for the 0.33 to 0.5 probability of resolution region. Comparisons are carried out in Fig. 3.3 using (3.85), (3.86) and simulation results from Table 3.1 for 0.33 to 0.5 probability of resolution. Fig. 3.4 show a similar comparison for yet another array length. In all these cases the close agreement between the theory and simulation results is clearly evident.

The above range (0.33 to 0.5) for the probability of resolution can be explained by reexamining the arguments used in deriving the resolution thresholds (3.85) and (3.86). In fact, (3.84) has been justified by observing that $Var[\hat{Q}(\omega_1)] = Var[\hat{Q}(\omega_2)] = 0$. Although $Var[\hat{Q}((\omega_1+\omega_2)/2)]$ is equally important in that analysis, it is nevertheless nonzero (see (3.C.43) and (3.C.44)). This implies that though ξ_T and ξ_T satisfy (3.84), in an actual set of trials the estimated mean value of $\hat{Q}[(\omega_1+\omega_2)/2]$ will almost always be inside the interval $(0, 2E[\hat{Q}((\omega_1+\omega_2)/2)])$ and clearly resolution of the two nulls in $\hat{Q}(\omega)$ is possible only if this mean estimate lies in the upper half of the above interval. In the special case of a symmetrical density function

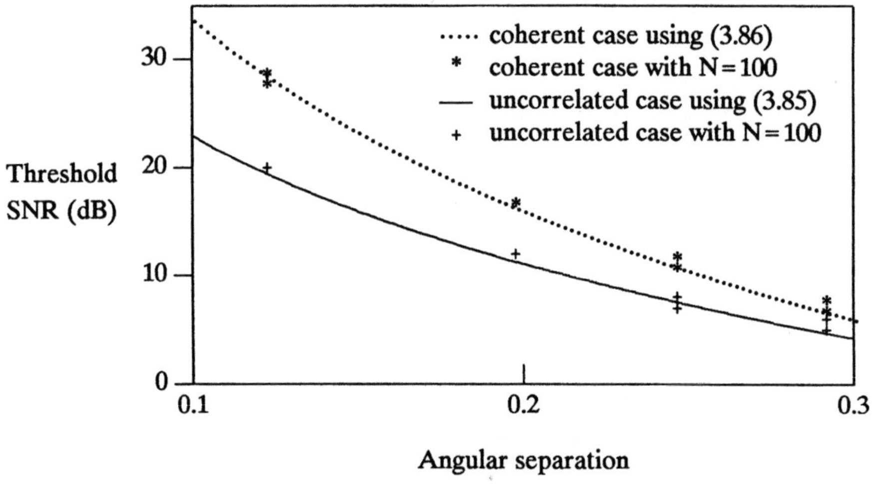

Fig. 3.3 Resolution threshold *vs.* angular separation for two equipowered sources in coherent and uncorrelated scenes. A seven-element array is used to receive signals in both cases.

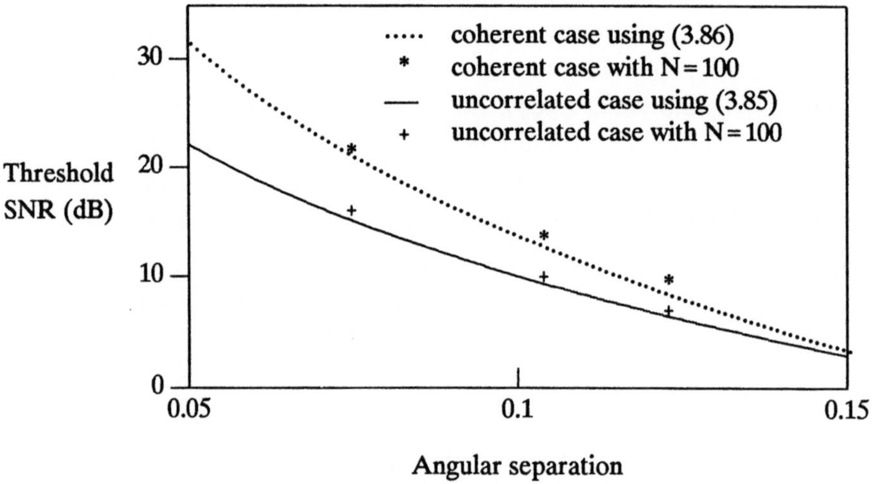

Fig. 3.4 Resolution threshold *vs.* angular separation for two equipowered sources in coherent and uncorrelated scenes. A fifteen-element array is used to receive signals in both cases.

for the above mean value estimate, this occurs with probability 0.5 and the observed range may be attributed to the skewed nature of the actual probability density function.

3.4 Performance Evaluation of GEESE Scheme

In this section, we examine the statistical behavior of the GEESE technique outlined in section 2.4.2 in a single source case and a two source case when estimated generalized eigenvalues are employed in its implementation. Referring back to (2.175) – (2.176), the row size J of the submatrices $\tilde{\mathbf{B}}_1$ and $\tilde{\mathbf{B}}_2$ can be any integer value between K and $M-1$. Clearly $J=K$ makes use of the least amount of information in terms of signal subspace eigenvectors and $J=M-1$ makes use of all available information in that respect. From here onwards we will refer to these extreme cases as least favorable and most favorable configurations respectively.

3.4.1 The Least Favorable Configuration (J = K)

Single Source Scene

In a single source scene $(K = 1)$, the eigenvector corresponding to the largest eigenvalue of the array output covariance matrix is

$$\beta_1 = \frac{1}{\sqrt{M}} [1, \nu_1, \cdots, \nu_1^{M-1}]$$

with ν_i, $i = 1, 2, \cdots, K$ as in (2.110) and from (2.178) – (2.179) since $\tilde{\mathbf{B}}_1$ and $\tilde{\mathbf{B}}_2$ are scalar quantities,

$$\hat{\gamma}_1 = \frac{c_{11}}{c_{21}} = \frac{\left(\beta_{11} + \dfrac{1}{\sqrt{N}} \displaystyle\sum_{\substack{j=1 \\ j \neq 1}}^{M} w_{j1}\beta_{1j}\right)}{\left(\beta_{21}\left(1 + \dfrac{1}{\sqrt{N}} \displaystyle\sum_{\substack{j=1 \\ j \neq 1}}^{M} w_{j1}\beta_{2j}/\beta_{21}\right)\right)} \tag{3.87}$$

where the second equality follows from (3.61) with $L = 1$

(unsmoothed case). For $|\frac{1}{\sqrt{N}} \sum\limits_{\substack{j=1 \\ j \neq 1}}^{M} w_{j1}\beta_{2j}/\beta_{21}| \ll 1$, the above ex-

pression simplifies into

$$\hat{\gamma}_1 = \gamma_1 + \frac{1}{\sqrt{N}}\Gamma_{11} + \frac{1}{N}\Gamma_{21} + \frac{1}{N\sqrt{N}}\Gamma_{31} + o\,(1/N^2) \qquad (3.88)$$

where

$$\Gamma_{11} = \sum\limits_{\substack{j=1 \\ j \neq 1}}^{M} w_{j1}\left[\beta_{1j}/\beta_{21} - \beta_{11}\beta_{2j}/\beta_{21}^2\right]$$

$$\Gamma_{21} = \sum\limits_{\substack{i=1 \\ i \neq 1}}^{M} \sum\limits_{\substack{j=1 \\ j \neq 1}}^{M} w_{i1}w_{j2}\left[\beta_{11}\beta_{2i}\beta_{2j}/\beta_{21}^3 - \beta_{1i}\beta_{2j}/\beta_{21}^2\right]$$

and

$$\Gamma_{31} = \sum\limits_{\substack{k=1 \\ j \neq 1}}^{M} \sum\limits_{\substack{i=1 \\ i \neq 1}}^{M} \sum\limits_{\substack{j=1 \\ j \neq 1}}^{M} w_{k1}w_{i1}w_{j1}\beta_{1k}\beta_{2i}\beta_{2j}/\beta_{21}^3 .$$

Since asymptotically the joint distribution of w_{ij}, $i = 1, 2, \cdots, K$, $j = 1, 2, \cdots, M$ tend to be normal with zero mean, their odd order moments are zeros and hence

$$E(\hat{\gamma}_1) = \gamma_1 + \frac{1}{N}\sum\limits_{\substack{i=1 \\ i \neq 1}}^{M} \sum\limits_{\substack{j=1 \\ j \neq 1}}^{M} E\left[w_{i1}w_{j1}\right]\left[\beta_{11}\beta_{2i}\beta_{2j}/\beta_{21}^3 - \beta_{1i}\beta_{2j}/\beta_{21}^2\right]$$

$$+ o\,(1/N^2) . \qquad (3.89)$$

Using (3.59) and (3.44) in (3.46) for $L = 1$, we get

$$E(w_{ij}w_{kl}) = \frac{\lambda_i\lambda_j}{(\lambda_j - \lambda_i)^2}\delta_{il}\delta_{jk} , \quad i \neq j \text{ and } k \neq l$$

and with this, (3.88) reduces to

$$E(\hat{\gamma}_1) = \gamma_1 + o\,(1/N^2) , \qquad (3.90)$$

which is an unbiased estimator for γ_1.

Further, with the help of (3.88) and (3.90), we also have

$$Var(\hat{\gamma}_1) = \frac{1}{N}E\left[\ |\ \frac{1}{\sqrt{N}}\Gamma_{11} + \frac{1}{N}\Gamma_{21} + \frac{1}{N\sqrt{N}}\Gamma_{31}\ |^2\right]$$

$$= \frac{1}{N}\sum_{\substack{i=1 \\ i \neq 1}}^{M}\sum_{\substack{j=1 \\ j \neq 1}}^{M}\left(E\left[w_{i1}w_{j1}^{*}\right]\left[\beta_{1i}/\beta_{21} - \beta_{11}\beta_{2i}/\beta_{21}^2\right]\right.$$

$$\left.\left[\beta_{1j}/\beta_{21} - \beta_{11}\beta_{2j}/\beta_{21}^2\right]^{*}\right) + o\,(1/N^2).$$

Since

$$E\,(w_{ij}w_{kl}^{*}) = \frac{\lambda_i\lambda_j}{(\lambda_i - \lambda_j)^2}\delta_{ik}\,\delta_{jl}\ ,\ \ i \neq j \text{ and } k \neq l$$

we have

$$Var(\hat{\gamma}_1) = \frac{\lambda_1\sigma^2}{N\,(\lambda_1 - \sigma^2)^2}\sum_{k=2}^{M}\left[\,|\,\beta_{1k}\,|^2 + |\,\beta_{2k}\,|^2 - 2\mathrm{Re}(\nu_1^{*}\beta_{1k}^{*}\beta_{2k})\right]$$

$$+ o\,(1/N^2)$$

which after some algebra reduces to

$$Var(\hat{\gamma}_1) = \frac{2M\lambda_1\sigma^2}{N\,(\lambda_1 - \sigma^2)^2} + o\,(1/N^2)$$

$$= \frac{2M}{N}\left[\frac{1}{\xi} + \frac{1}{\xi^2}\right] + o\,(1/N^2) \tag{3.91}$$

where we have made use of $\lambda_1 = MP + \sigma^2$. Here $\xi = MP/\sigma^2$ represents the array output signal-to-noise ratio.

Two Source Scene

With the help of (3.61) together with (3.C.27), after considerable manipulations the mean and variance of the estimated generalized eigenvalues $\hat{\gamma}_i$, $i = 1, 2$ in a two equipowered uncorrelated source scene can be shown to be [34, 35]

$$E[\hat{\gamma}_i] = \gamma_i + o\left(\frac{1}{N\sqrt{N}}\right), \quad i = 1, 2, \tag{3.92}$$

and

$$Var(\hat{\gamma}_i) = \frac{M\left(2 + \text{Re}(\nu_1\nu_2^*)\right)}{2N\left(1 - \text{Re}(\nu_1\nu_2^*)\right)}\left[(1 + |\rho_s|)\frac{\lambda_1\sigma^2}{(\lambda_1 - \sigma^2)^2}\right.$$

$$\left. + (1 - |\rho_s|)\frac{\lambda_2\sigma^2}{(\lambda_2 - \sigma^2)^2}\right] + o\left(\frac{1}{N\sqrt{N}}\right), \quad i = 1, 2 \tag{3.93}$$

where ρ_s is as defined in (3.C.9). Thus, for $J = K = 2$, within a $1/N$ approximation, $\hat{\gamma}_i$, $i = 1, 2$ are once again unbiased estimates with finite variance. Simulation results presented in Fig. 3.5 seem to be in agreement with these conclusions. The random pattern for actual bias in Fig. 3.5 may be attributed to computational and other round off errors and indicates the absence of a $1/N$ term there.

To simplify (3.93) further in an equipowered uncorrelated two source scene, (3.C.26) can be used to obtain the signal subspace eigenvalues. This gives

$$\lambda_i = MP(1 \pm |\rho_s|) + \sigma^2, \quad i = 1, 2. \tag{3.94}$$

With (3.94) in (3.93), finally it simplifies into the convenient form

$$Var(\hat{\gamma}_i) = \frac{M(2 + \cos 2\omega_d)}{N(1 - \cos 2\omega_d)}\left[\frac{1}{\xi} + \frac{1}{\xi^2(1 - |\rho_s|^2)}\right]$$

$$+ o\left(\frac{1}{N\sqrt{N}}\right), \quad i = 1, 2. \tag{3.95}$$

These expressions can be used to determine the resolution threshold associated with two closely spaced sources. As remarked earlier for a specific input SNR, the resolution threshold represents the

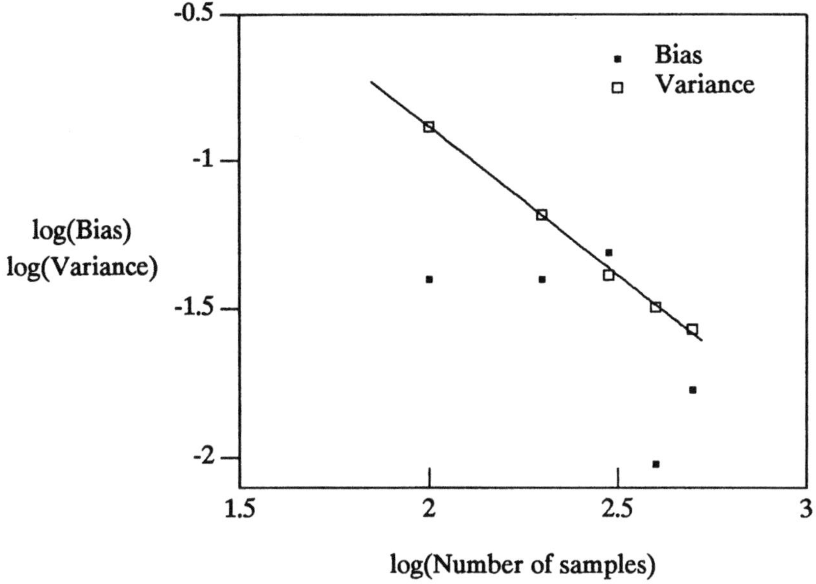

Fig. 3.5 Bias and Variance for the least favorable configuration ($J = K$ ($=2$)). Bias and variance *vs.* number of snapshots for two equipowered sources. A ten-element array is used to receive signals from two uncorrelated sources located along $45°$, $50°$ with common SNR $=$ 5dB. Each simulation here consists of 100 independent trials.

minimum amount of angular separation required to identify the sources as separate entities unambiguously. From (3.92) and (3.95), since the standard derivation of $\hat{\gamma}_i$, $i = 1, 2$ is substantially larger than their respective bias, it is clear that the resolution threshold is mostly determined by the behavior of the standard deviation. In order to obtain a measure of the resolution threshold for two closely spaced sources, consider the situation shown in Fig. 3.6. Evidently, the sources are resolvable if $\hat{\gamma}_1$ and $\hat{\gamma}_2$ are both inside the cones C_1 and C_2 respectively or equivalently if $|\, arg\,(\hat{\gamma}_i) - arg\,(\gamma_i)\,| < \omega_d$, $i = 1, 2$. Exact calculations based on this criteria turn out to be rather tedious. But as computation results in Fig. 2.16 show, $\hat{\gamma}_1$ and $\hat{\gamma}_2$ are usually within a small circular neighborhood centered about γ_1 and γ_2. This suggests a more conservative criterion for resolution; i.e., the sources are resolvable if $\hat{\gamma}_1$ and $\hat{\gamma}_2$ are both inside the circles c_1 and c_2 respec-

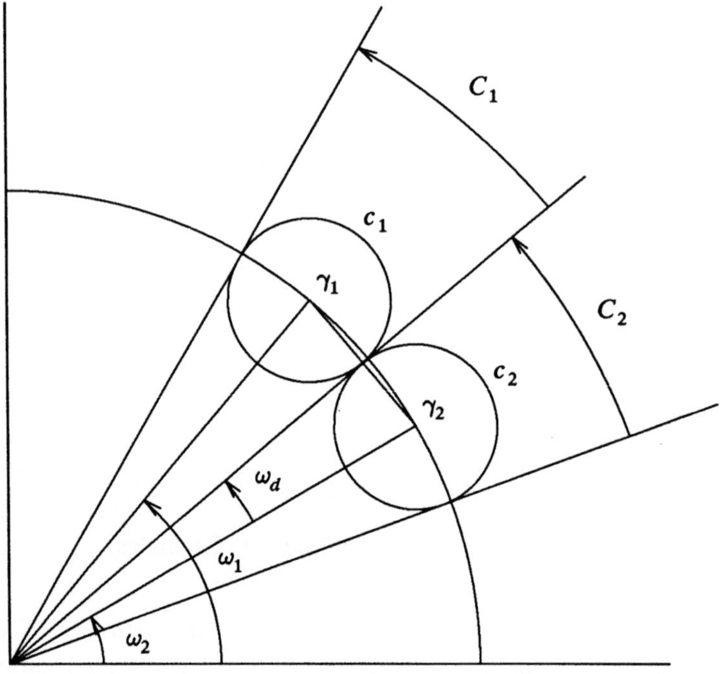

Fig. 3.6 Resolution threshold analysis.

tively in Fig. 3.6. In that case, the maximum value of the common radii of these circles is easily shown to be $\sin \omega_d$. Thus, at an SNR satisfying

$$\sqrt{Var(\hat{\gamma}_i)} = l \sin \omega_d \tag{3.96}$$

where l is some positive integer, using (3.96) the associated threshold SNR turns out to be

$$\xi_{l,K} = \frac{M(2 + \cos 2\omega_d)}{2l^2 N(1 - \cos 2\omega_d)\sin^2 \omega_d}$$

$$\times \left[1 + \left(1 + \frac{4l^2 N(1 - \cos 2\omega_d)\sin^2 \omega_d}{M(2 + \cos 2\omega_d)(1 - |\rho_s|^2)} \right)^{1/2} \right]. \tag{3.97}$$

This threshold SNR can also be expressed in terms of the "effective

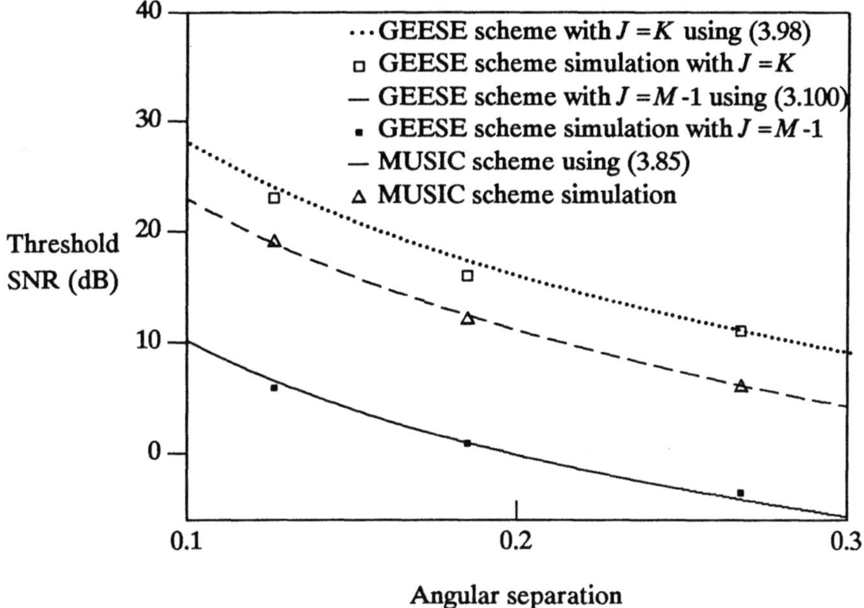

Fig. 3.7 Resolution threshold *vs.* angular separation for two equipowered sources. A seven-element array is used to receive signals in all cases. One hundred snapshots are taken for each simulation. In each simulation, the associated probability of resolution is 30 percent. The threshold SNRs of GEESE scheme are calculated with $l=2$ in both (3.98) and (3.100).

angular separation" parameter Δ^2 defined in (3.C.28). For closely spaced sources, $M^2 \omega_d^2 << 1$ and in that case using (3.C.31) we have

$$\xi_{l,K} \simeq \frac{M}{6l^2N}\left(\frac{M^4}{2\Delta^4} - \frac{M^2}{\Delta^2}\right)\left[1 + \left(1 + \frac{24l^2N\Delta^2}{M^5}\right)^{1/2}\right] \quad (3.98)$$

Notice that calculations for Var $(\hat{\gamma}_i)$ in (3.95) have been carried out for $J = K(=2)$ case, and hence the above threshold expression also corresponds to this least favorable configuration, which only uses part of the available signal subspace eigenvector information in its computations. When higher value of J is used to evaluate $\hat{\gamma}_i$, the corresponding threshold expressions also should turn out to be superior to that in (3.98). These conclusions are seen to closely agree with results of

Table 3.2 Resolution threshold and probability of resolution *vs.* angular separation for two equipowered sources ($K = 2$) in an uncorrelated scene (number of sensors M = 7, number of snapshots = 100, number of simulations = 100).

angles of arrival θ_1	θ_2	angular separation $2\omega_d$	dB	MUSIC	GEESE $J=K$	$J=4$	$J=M-1$
			4				0.14
			6				0.33
			8				0.54
			16		0.12	0.64	0.98
25.00	35.00	0.1265	18	0.19	0.19	0.77	1.00
			19	0.31	0.20	0.84	
			20	0.42	0.20	0.89	
			22	0.71	0.25	0.98	
			24	0.91	0.40		
			0				0.27
			1				0.31
			3				0.57
			9		0.09	0.58	0.99
40.00	45.00	0.1852	11	0.12	0.16	0.77	1.00
			13	0.32	0.21	0.91	
			14	0.48	0.23	0.94	
			15	0.73	0.26	0.97	
			17	0.84	0.31	1.00	
			-4				0.23
			-2				0.51
			0			0.28	0.66
			4		0.08	0.66	0.98
75.00	80.00	0.2676	6	0.29	0.11	0.85	1.00
			7	0.41	0.22	0.94	
			8	0.57	0.16	0.92	
			10	0.81	0.25	0.97	
			12	0.99	0.56		

simulation presented in Table 3.2. Similar threshold comparisons are carried out in Fig. 3.7 for the MUSIC scheme using (3.85) and the GEESE scheme using (3.98) with $l = 2$. From Table 3.2, the corresponding SNR values are observed to have at least 30 percent probability of resolution.

3.4.2 The Most Favorable Configuration (J = M − 1)

For $J > K$, the situation is much more complex and the estimator $\hat{\gamma}_i$, $i = 1, 2$ turns out to be no longer unbiased even within a $1/N$ approximation. The exact bias and variance expressions have been computed for the most favorable configuration ($J = M - 1$) in a two source scene [35]. In particular, it can be shown that

$$Var(\hat{\gamma}_i) = \frac{c(M,\Delta^2)}{N} \left[\frac{\Delta^2}{\xi} + \frac{1}{\xi^2} \right], \quad i = 1, 2 \tag{3.99}$$

where

$$c(M,\Delta^2) = \frac{2M^5}{(M-1)^6} \left[\left(3 - \frac{2}{M^2} \right) \frac{1}{\Delta^4} + \left(\frac{2}{5} - \frac{14}{M} + \frac{123}{M^2} \right) \frac{1}{\Delta^2} \right.$$
$$\left. - \left(2 - \frac{7}{M} - \frac{15}{M^2} + \frac{95}{M^3} \right) \right].$$

The $1/N$ dependence for bias and variance is also evident in the simulation results presented in Fig. 3.8. The associated threshold SNR in this case can be obtained with the help of (3.96) and (3.99) and this gives

$$\xi_{l,M-1} \simeq \frac{M^2 c(M,\Delta^2)}{6l^2 N} \left[1 + \left(1 + \frac{2l^2 N (M-1)^6 \Delta^2}{M^5(M^2-2)} \right)^{1/2} \right] \tag{3.100}$$

For the same source scene and probability of resolution discussed in Fig. 3.7, new simulation results are also presented there for this most favorable configuration. As remarked earlier, the SNR required to resolve two uncorrelated sources in this case is seen to be substantially smaller than that in the former case ($J = K$). Once again, utilization of all available information in this case ($J = M - 1$) may be attributed to its superior performance. From these results, it

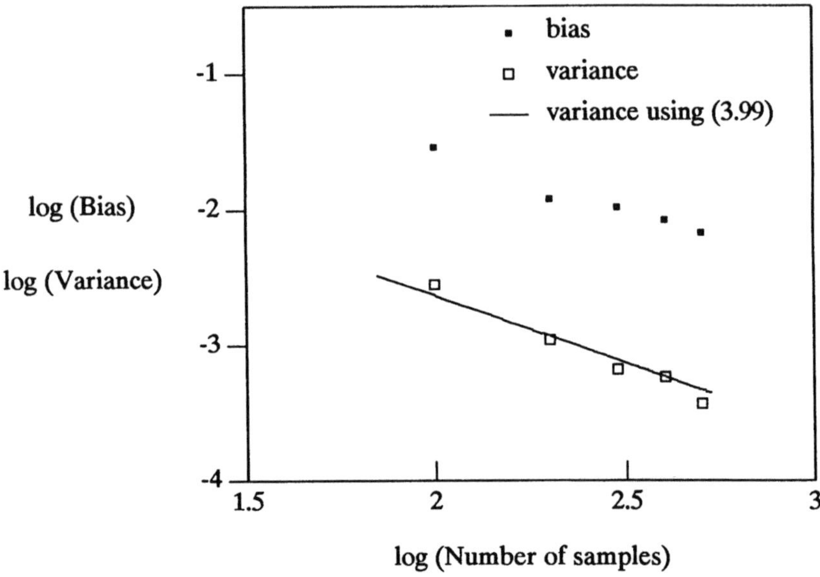

Fig. 3.8 Bias and Variance for the most favorable configuration $(J = M - 1)$. Bias and variance *vs.* number of snapshots for two equipowered sources. A ten element array is used to receive signals from two uncorrelated sources located along 30°, 35° with common SNR = 5dB. Each simulation here consists of 100 independent trials.

may be reasonably concluded that when all available signal subspace information is exploited, the GEESE scheme outperforms the MUSIC scheme.

3.5 Estimation of Number of Signals

So far we have proceeded under the assumption that the total number of signals present in the data is known. In practice, that number K is unknown and it also will have to be estimated from the data. Referring back to the signal model (2.53) in section 2.3, in presence of independent and identically distributed noise, the structural properties associated with the array output covariance matrix $\mathbf{R}$ can be readily exploited to solve this problem. In fact the equality of the $(M - K)$ lowest eigenvalues of $\mathbf{R}$ in (2.60) reveals that the problem at hand is trivial, given the ensemble average of this covariance matrix. However, when only an estimate $\mathbf{S}$ of $\mathbf{R}$ as in (3.7) is available, this is no longer a trivial problem, since with probability 1, none of the eigenvalues of $\mathbf{S}$ will be equal [12]. Statistical tests based on the asymptotic multivariate Gaussian nature of the eigenvalues, have been devised for the real data case by Anderson [13], Lawley [14], and others [15], to estimate the number of equal eigenvalues with a certain confidence level. Similar results for the complex case can be found in [16, 17]. Its use in array processing is reported in [18, 19] and extension of these methods to coherent scenarios is discussed in [20, 21]. As before, let

$$\mathbf{S} = \mathbf{ELE}^{\dagger}, \quad \mathbf{EE}^{\dagger} = \mathbf{I}$$

and

$$\mathbf{L} = diag\,[\,l_1, l_2, \cdots, l_M\,]$$

such that $l_1 \geq l_2 \geq \cdots \geq l_M$ represent the ordered eigenvalues of the ML estimate $\mathbf{S}$. The likelihood ratio criterion for the equality of the $(M-K)$ lowest eigenvalues in the complex case has been shown to be [17]

$$\delta_K(l) \triangleq \left(\frac{1}{M-K} \sum_{i=K+1}^{M} l_i \right) \Bigg/ \left(\prod_{i=K+1}^{M} l_i \right)^{\frac{1}{M-K}} \tag{3.101}$$

which is the ratio of the arithmetic mean to the geometric mean of the corresponding sample eigenvalues. Using an asymptotic analysis, it can be shown that $2N(M-K)\ln\delta_K(l)$ is a χ^2-distributed random variable, i.e.,

$$2N(M-K)\ln\delta_K(l) \sim \chi^2(m_0) \tag{3.102}$$

where

$$m_0 = (M-K)^2 - 1 \tag{3.103}$$

represents the degree of freedom associated with the χ^2 random variable in (3.102). This will allow one to examine the equality of the lowest $(M-K)$ eigenvalues by computing $2N(M-K)\ln\delta_K(l)$ and comparing it with the upper tail significance point of the appropriate χ^2-distribution for some suitable significance level α. In this comparison, if the computed quantity turns out to be less than the upper tail significance point, then the equality of the $M-K$ lowest eigenvalues is accepted. The test will proceed with $K = 1, 2, \cdots$ and the first value of K for which the test succeeds is taken as the estimate for K [18 - 20]. Clearly the test is meaningful only if $K < M$, i.e., so long as the number of targets is less than the number of sensor elements. In what follows, instead of proving (3.102) we derive an alternate test based on higher powers of the eigenvalues and derive its asymptotic distribution.

With $\lambda_1, \lambda_2, \cdots, \lambda_M$ representing the eigenvalues of $\mathbf{R}$, the hypothesis H_K under test is $\lambda_{K+1} = \lambda_{K+2} = \lambda_M$. The desired criterion for the equality of the $(M-K)$ lowest eigenvalues of $\mathbf{S}$, based on their p^{th} powers is given by

$$
\Delta_K(l,p) = \left[\left(\frac{1}{M-K} \sum_{i=K+1}^{M} l_i^p \right) \Big/ \left(\prod_{i=K+1}^{M} l_i^p \right)^{\frac{1}{M-K}} \right]^{1/p}
$$

$$
= \left(\frac{1}{M-K} \sum_{i=K+1}^{M} l_i^p \right)^{1/p} \Big/ \left(\prod_{i=K+1}^{M} l_i \right)^{\frac{1}{M-K}} . \qquad (3.104)
$$

We will show that

$$
\frac{2N(M-K)}{p} \ln \Delta_K(l,p) \sim \chi^2[(M-K)^2-1] . \qquad (3.105)
$$

Proof:

Following Anderson [4], define

$$
d_i = \sqrt{N} \, (l_i - \lambda_{K+1}), \quad i = K+1, K+2, M .
$$

This gives

$$
l_i = \frac{1}{\sqrt{N}} d_i + \lambda_{K+1} . \qquad (3.106)
$$

With (3.104) and (3.106) in (3.105) we have

$$\frac{2N(M-K)}{p}\, ln\, \Delta_K(l,p)$$

$$= \frac{2N}{p}\left[\frac{(M-K)}{p}\, ln\left\{\frac{1}{(M-K)}\sum_{i=K+1}^{M}\left(1+\frac{N^{-1/2}d_i}{\lambda_{K+1}}\right)^p\right\}\right.$$

$$\left. - \sum_{i=K+1}^{M} ln\left(1+\frac{N^{-1/2}d_i}{\lambda_{K+1}}\right)\right]$$

$$= \frac{2N}{p}\left[\frac{(M-K)}{p}\, ln\left\{1+\left(\frac{p\sum d_i}{\sqrt{N}(M-K)\lambda_{K+1}}+\frac{p(p-1)\sum d_i^2}{2N(M-K)\lambda_{K+1}^2}+\cdots\right)\right\}\right.$$

$$\left. -\left(\frac{\sum d_i}{\sqrt{N}\,\lambda_{K+1}}-\frac{\sum d_i^2}{2N\,\lambda_{K+1}^2}+\cdots\right)\right]$$

$$= \frac{2N}{p}\left[\frac{(M-K)}{p}\left(\frac{p\sum d_i}{\sqrt{N}(M-K)\lambda_{K+1}}+\frac{p(p-1)\sum d_i^2}{2N(M-K)\lambda_{K+1}^2}\right.\right.$$

$$\left.\left. -\frac{p^2(\sum d_i)^2}{2N(M-K)^2\lambda_{K+1}^2}+\cdots\right)-\frac{\sum d_i}{\sqrt{N}\,\lambda_{K+1}}+\frac{\sum d_i^2}{2N\,\lambda_{K+1}^2}-\cdots\right]$$

$$= \frac{1}{\lambda_{K+1}^2}\left[\sum_{i=K+1}^{M}d_i^2-\frac{1}{(M-K)}\left(\sum_{i=K+1}^{M}d_i\right)^2\right]+o(1/N\sqrt{N}). \quad (3.107)$$

For $L=1$, since $d_i = f_{ii}$, $i=1, 2, \cdots, M$ in (3.31), using that together with (3.33) and (3.34), it follows that the limiting distribution of $d_{K+1}, d_{K+2}, \cdots, d_M$, is the same as the distribution of the eigenvalues of the hermitian matrix $\mathbf{U}_{22} = (u_{ij})$, $i, j = K+1, K+2, \cdots, M$ in (3.49). Further, for circular Gaussian data, with (3.44) in (3.43) (with $L = 1$) we have $E[u_{ij} u_{kl}^*] = \lambda_i \lambda_j \delta_{ik} \delta_{jl}$, i.e., elements of $\mathbf{U}_{22}$ are asymptotically independent, identically distributed, zero mean Gaussian random variables with common variance λ_{K+1}^2. Then it follows that (3.107) has the limiting distribution of

$$\frac{1}{\lambda_{K+1}^2} \left[tr(\mathbf{U}_{22}^2) - \frac{1}{(M-K)} \left(tr(\mathbf{U}_{22}) \right)^2 \right]$$

$$= \frac{1}{\lambda_{K+1}^2} \left[tr\left(\mathbf{U}_{22} \mathbf{U}_{22}^\dagger \right) - \frac{1}{(M-K)} \left(tr(\mathbf{U}_{22}) \right)^2 \right]$$

$$= \frac{1}{\lambda_{K+1}^2} \left[\left(\sum_{i \neq j} |u_{ij}|^2 + \sum_{i=K+1}^{M} u_{ii}^2 \right) - \frac{1}{(M-K)} \left(\sum_{i=K+1}^{M} u_{ii} \right)^2 \right]$$

$$= \frac{1}{\lambda_{K+1}^2} \sum_{i \neq j} |u_{ij}|^2 + \frac{1}{\lambda_{K+1}^2} \left[\sum_{i=K+1}^{M} u_{ii}^2 - \frac{1}{(M-K)} \left(\sum_{i=K+1}^{M} u_{ii} \right)^2 \right].$$

$$(3.108)$$

Thus $\left(\sum_{i \neq j} |u_{ij}|^2 \right)/\lambda_{K+1}^2$ is asymptotically χ^2 with $(M-K)(M-K-1)$ degrees of freedom; independently, $[\sum_i u_{ii}^2 - (\sum_i u_{ii})^2/(M-K)]/\lambda_{K+1}^2$ is asymptotically χ^2 with $(M-K-1)$ degrees of freedom [2]. Thus (3.107) has a limiting χ^2-distribution with $[(M-K)^2-1]$ degrees of freedom and hence

$$\frac{2N(M-K)}{p} \ln \Delta_K(l,p) \sim \chi^2[(M-K)^2-1].$$

This completes the proof. Clearly, the test in (3.102) based on the optimum likelihood ratio criterion (3.101) follows as a special case of (3.105) with $p = 1$.

Although the test in (3.101) is designed to satisfy the likelihood criterion and is optimal in that sense, nevertheless to utilize it in practice, the significance level α has to be chosen subjectively (refer to the discussion after (3.103)). Since the significance level alters the estimated number of signals, this serious difficulty can be avoided only by reformulating the problem, and several solutions have been reported in this direction. Akaike's Information Criterion (AIC) [22] and the Minimum Description Length (MDL) proposed by Rissanen [23] based on Kullback-Leibler information measure [24] are solutions that avoid the difficulties involved with subjective decisions. Wax and Kailath [25] have extended the use of AIC and MDL criteria for the estimation of number of signals. According to the AIC criterion, the model for which

$$AIC(K) = -2N(M-K)ln\delta_K(l) + 2\nu(K,M) \qquad (3.109)$$

is minimum is selected as K varies from 0 to M-1. Here $\nu(K,M)$ denotes the number of free parameters that have to be estimated under the hypothesis H_K. To begin with, there are $K+1$ real eigenvalues and K complex eigenvectors (associated with λ_1, $\lambda_2 \cdots, \lambda_K$) which together give rise to $K+1+2MK$ free parameters. Notice that the eigenvectors associated with σ^2 have no restrictions and can be chosen orthogonal to those in the signal subspace. However, not all of the above parameters can be independently adjusted. In fact the unit norm constraint and their mutual orthogonality reduces the number of free parameters by $2K$ and $2[K(K-1)/2]$ respectively. Put together, this gives the number of free parameters to be

$$\nu(K,M) = K+1+2MK-2K-K(K-1) = K(2M-K)+1 \qquad (3.110)$$

The MDL criterion, on the other hand, selects the model for which

$$MDL(K) = -2N(M-K)ln\delta_K(l) + 2\nu(K,M)\frac{ln N}{2} \qquad (3.111)$$

is minimum. Within this framework of model selection, Zhao *et al.* [26] have proposed a new procedure for estimation of the number of signals. This procedure, known as the Efficient Detection Criterion (EDC) computes

$$EDC(K) = -2N(M-K)\ln\delta_K(l) + 2\nu(K,M)C(N) \quad (3.112)$$

for $K = 0, 1, \cdots, M$-1 and selects the model that minimizes the above function. In (3.112) if $C(N)$ is chosen such that it satisfies

$$\lim_{N\to\infty} C(N)/N = 0 \quad (3.113)$$

and

$$\lim_{N\to\infty} C(N)/\ln(\ln N) = \infty \quad (3.114)$$

then the strong consistency of the above procedure can be established [26]. Notice that MDL criterion is a special case of EDC criterion with $C(N) = (\ln N)/2$, and this establishes the strong consistency of the MDL criterion. A generalization of these procedures based on r-regular functions is reported in [27] for estimating the number of signals present in the scene.

Appendix 3.A
The Complex Wishart Distribution

Let $\mathbf{x}(n)$ represent an independent sample of N observations from an M-variate zero mean complex Gaussian random vectors with covariance matrix $\mathbf{R}$. (i.e., $\mathbf{x}(n) \sim N(\mathbf{0},\mathbf{R})$. Then

$$\mathbf{S} = \frac{1}{N}\mathbf{B} \quad (3.A.1)$$

with

$$\mathbf{B} = \sum_{n=1}^{N} \mathbf{x}(n)\mathbf{x}^\dagger(n) \quad (3.A.2)$$

is the maximum likelihood (ML) estimator of the unknown covariance matrix $\mathbf{R}$, and the joint probability density function (p.d.f) of the functionally independent elements $B_{ij}, i\leq j = 1,2,\cdots,M$ is known as the complex Wishart distribution. The characteristic function of the complex Wishart distribution can be obtained directly from the observations. In fact, for any $M\times M$ hermitian matrix Θ, it is given by

$$E(e^{j tr(\mathbf{B}\Theta)}) = E\left[e^{j\, tr(\sum_{n=1}^{N} \mathbf{x}(n)\mathbf{x}^\dagger(n)\Theta)}\right] = E\left[e^{j\, tr(\sum_{n=1}^{N} \mathbf{x}^\dagger(n)\Theta\mathbf{x}(n))}\right]$$

$$= \prod_{n=1}^{N} E\left[e^{j\sum_{n=1}^{N} \mathbf{x}^\dagger(n)\Theta\mathbf{x}(n)}\right] = \left[E\left(e^{j\sum_{n=1}^{N} \mathbf{x}^\dagger(n)\Theta\mathbf{x}(n)}\right)\right]^{N}. \qquad (3.A.3)$$

To simplify this further, notice that for Θ hermitian, there exists a non-singular matrix $\mathbf{M}$ such that

$$\mathbf{M}^\dagger\mathbf{R}^{-1}\mathbf{M} = \mathbf{I} \quad \text{and} \quad \mathbf{M}^\dagger\Theta\mathbf{M} = \mathbf{D}$$

where $\mathbf{D}$ is a real diagonal matrix. Let

$$\xi = \mathbf{M}^{-1}\mathbf{x}(n), \quad \text{then } E[\xi\xi^\dagger] = \mathbf{I}$$

and

$$E(e^{j\mathbf{x}^\dagger(n)\Theta\mathbf{x}(n)}) = E(e^{j\xi^\dagger\mathbf{M}^\dagger\Theta\mathbf{M}\xi}) = E(e^{j\xi^\dagger\mathbf{D}\xi})$$

$$= \prod_{i=1}^{M} E(e^{jD_{ii}|\xi_i|^2}) = \prod_{i=1}^{M} (1 - j D_{ii})^{-1} = |\,\mathbf{I} - j\,\mathbf{D}\,|^{-1}$$

$$= |\,\mathbf{M}^\dagger\mathbf{R}^{-1}\mathbf{M} - j\,\mathbf{M}^\dagger\Theta\mathbf{M}\,|^{-1}$$

$$= |\,\mathbf{M}^\dagger\,|^{-1}\,|\,\mathbf{I} - j\,\Theta\mathbf{R}\,|^{-1}\,|\,\mathbf{R}^{-1}\mathbf{M}\,|^{-1} = |\,\mathbf{I} - j\,\Theta\mathbf{R}\,|^{-1}.$$

Thus from (3.A.3), the desired characteristic function is given by

$$E\left[e^{j\, tr(\mathbf{B}\Theta)}\right] = |\,\mathbf{I} - j\,\Theta\mathbf{R}\,|^{-N}. \qquad (3.A.4)$$

Goodman has shown the p.d.f. corresponding to (3.A.4) to be [5]

$$f_B(\mathbf{B}) = \frac{|\,\mathbf{B}\,|^{N-M}}{\Gamma_M(N)\,|\,\mathbf{R}\,|^{N}}\, e^{-tr(\mathbf{R}^{-1}\mathbf{B})} \qquad (3.A.5)$$

where

$$\Gamma_M(N) = \pi^{M(M-1)/2}\Gamma(N)\Gamma(N-1)\cdots\Gamma(N-M+1).$$

This joint p.d.f is denoted by $CW(N,M,\mathbf{R})$ and is defined over the region where $\mathbf{B}$ is at least nonnegative definite. From (3.A.5) we also have the useful identity

$$\int |\mathbf{B}|^{N-M} e^{-tr(\mathbf{R}^{-1}\mathbf{B})} d\mathbf{B} = \Gamma_M(N) |\mathbf{R}|^N.$$

The Sum of Wishart Matrices

Let $\mathbf{B}_1 \sim CW(N_1,M,\mathbf{R})$ and $\mathbf{B}_2 \sim CW(N_2,M,\mathbf{R})$ be independent. Then,

$$\mathbf{B}_1 + \mathbf{B}_2 \sim CW(N_1+N_2,M,\mathbf{R}) \qquad (3.A.6)$$

This follows easily by noticing that

$$\mathbf{B}_1 = \sum_{n=1}^{N_1} \mathbf{x}(n)\mathbf{x}^\dagger(n), \quad \mathbf{B}_2 = \sum_{n=1}^{N_2} \mathbf{x}(n)\mathbf{x}^\dagger(n)$$

where $\mathbf{x}(1), \mathbf{x}(2),..., \mathbf{x}(N_1+N_2)$ are i.i.d. random vectors, each distributed as $N(\mathbf{0},\mathbf{R})$. Hence by definition

$$\mathbf{B}_1 + \mathbf{B}_2 = \sum_{n=1}^{N_1+N_2} \mathbf{x}(n)\mathbf{x}^\dagger(n) \sim CW(N_1+N_2,M,\mathbf{R}).$$

A Certain Linear Transformation

If $\mathbf{B} \sim CW(N,M_1,\mathbf{R})$, then for any $M_2 \times M_1$ matrix $\mathbf{C}$,

$$\mathbf{C}\mathbf{B}\mathbf{C}^\dagger \sim CW(N,M_2,\mathbf{C}\mathbf{R}\mathbf{C}^\dagger). \qquad (3.A.7)$$

To see this, start with $\mathbf{B}$ as in (3.A.2). Then,

$$\mathbf{C}\mathbf{B}\mathbf{C}^\dagger = \sum_{n=1}^{N} (\mathbf{C}\mathbf{x}(n))(\mathbf{C}\mathbf{x}(n))^\dagger \sim CW(N,M_2,\mathbf{C}\mathbf{R}\mathbf{C}^\dagger)$$

since $\mathbf{C}\mathbf{x}(n) \sim N(\mathbf{0},\mathbf{C}\mathbf{R}\mathbf{C}^\dagger), n = 1,2,\cdots,N$ are independent random vectors.

The Schur Complements and Regression Coefficients

Let $\mathbf{B} \sim CW(N, M, \mathbf{R})$. If the positive definite matrices $\mathbf{B}$, $\mathbf{B}^{-1}$, $\mathbf{R}$ and $\mathbf{R}^{-1}$ are partitioned by the first M_1 and the rest $M_2 = M - M_1$ rows and columns in the form

$$
\mathbf{B} = \begin{bmatrix} \mathbf{B}_{11} & \mathbf{B}_{12} \\ \mathbf{B}_{21} & \mathbf{B}_{22} \end{bmatrix}, \quad
\mathbf{B}^{-1} = \begin{bmatrix} \mathbf{B}^{11} & \mathbf{B}^{12} \\ \mathbf{B}^{21} & \mathbf{B}^{22} \end{bmatrix}
$$

$$
\mathbf{R} = \begin{bmatrix} \mathbf{R}_{11} & \mathbf{R}_{12} \\ \mathbf{R}_{21} & \mathbf{R}_{22} \end{bmatrix}, \quad
\mathbf{R}^{-1} = \begin{bmatrix} \mathbf{R}^{11} & \mathbf{R}^{12} \\ \mathbf{R}^{21} & \mathbf{R}^{22} \end{bmatrix}.
$$

Then,

i) The $M_2 \times M_2$ Schur complement [3]

$$
(\mathbf{B}^{22})^{-1} = \mathbf{B}_{22} - \mathbf{B}_{21}\mathbf{B}_{11}^{-1}\mathbf{B}_{12} \sim CW(N - M_1, M_2, (\mathbf{R}^{22})^{-1}) \quad (3.A.8)
$$

where

$$
(\mathbf{R}^{22})^{-1} = \mathbf{R}_{22} - \mathbf{R}_{21}\mathbf{R}_{11}^{-1}\mathbf{R}_{12} .
$$

ii) The p.d.f of the $M_1 \times M_2$ regression coefficient $\mathbf{b} = \mathbf{B}_{11}^{-1}\mathbf{B}_{12}$ given $\mathbf{B}_{11}$, is conditionally Gaussian as

$$
f_{\mathbf{b}|\mathbf{B}_{11}}(\mathbf{b}|\mathbf{B}_{11}) = \frac{|\mathbf{B}_{11}|^{M_2}}{\pi^{M_1 M_2}|(\mathbf{R}^{22})^{-1}|^{M_1}} e^{-tr[(\mathbf{b}-\boldsymbol{\beta})^\dagger \mathbf{B}_{11}(\mathbf{b}-\boldsymbol{\beta})\mathbf{R}^{22}]} \quad (3.A.9)
$$

------

[3]

$$
\begin{bmatrix} \mathbf{A} & \mathbf{B} \\ \mathbf{B}^\dagger & \mathbf{D} \end{bmatrix}^{-1} = \begin{bmatrix} \mathbf{A}^{-1} + \mathbf{F}\mathbf{E}^{-1}\mathbf{F}^\dagger & -\mathbf{F}\mathbf{E}^\dagger \\ -\mathbf{E}\mathbf{F}^\dagger & \mathbf{E}^{-1} \end{bmatrix}
$$

where $\mathbf{E} = \mathbf{D} - \mathbf{B}^\dagger \mathbf{A}^{-1}\mathbf{B}$, $\mathbf{F} = \mathbf{A}^{-1}\mathbf{B}$.

where $\beta = \mathbf{R}_{11}^{-1}\mathbf{R}_{12}$.

iii) Also,

$$\mathbf{B}_{11} \sim CW(N, M, \mathbf{R}_{11}) \qquad (3.\mathrm{A}.10)$$

and more importantly $(\mathbf{B}^{22})^{-1}$ is statistically independent of $\mathbf{b}$ and $\mathbf{B}_{11}$.

To prove these standard results [28, 29], define the transformation of the matrix $\mathbf{B}$ as

$$\mathbf{X} = (\mathbf{B}^{22})^{-1} = \mathbf{B}_{22} - \mathbf{B}_{21}\mathbf{B}_{11}^{-1}\mathbf{B}_{12}, \quad \mathbf{Y} = \mathbf{B}_{11}, \quad \mathbf{Z} = \mathbf{B}_{11}^{-1}\mathbf{B}_{12}.$$

The inverse relations are clearly

$$\mathbf{B}_{11} = \mathbf{Y}, \quad \mathbf{B}_{12} = \mathbf{Y}\mathbf{Z}, \quad \mathbf{B}_{22} = \mathbf{X} + \mathbf{Z}^{\dagger}\mathbf{Y}\mathbf{Z}.$$

A direct computation yields the Jacobian of this complex transformation to be [28]

$$J(\mathbf{B};\mathbf{X},\mathbf{Y},\mathbf{Z}) = |\mathbf{Y}|^{2M_2}.$$

Applying this transformation,

$$f_{\mathbf{X},\mathbf{Y},\mathbf{Z}}(\mathbf{X},\mathbf{Y},\mathbf{Z}) = |J| \, f_B\begin{pmatrix} \mathbf{Y} & \mathbf{Y}\mathbf{Z} \\ (\mathbf{Y}\mathbf{Z})^{\dagger} & \mathbf{X} + \mathbf{Z}^{\dagger}\mathbf{Y}\mathbf{Z} \end{pmatrix}. \qquad (3.\mathrm{A}.11)$$

Noticing that

$$|\mathbf{B}| = |\mathbf{B}_{11}| \; |\mathbf{B}_{22} - \mathbf{B}_{21}\mathbf{B}_{11}^{-1}\mathbf{B}_{12}| = |\mathbf{Y}| \; |\mathbf{X}| \qquad (3.\mathrm{A}12)$$

$$|\mathbf{R}| = |\mathbf{R}_{11}| \; |\mathbf{R}_{22} - \mathbf{R}_{21}\mathbf{R}_{11}^{-1}\mathbf{R}_{12}| = |\mathbf{R}_{11}| \; |\mathbf{R}^{22}|^{-1} \qquad (3.\mathrm{A}13)$$

and with $^{(3)}$ $\mathbf{F} = \mathbf{R}_{11}^{-1}\mathbf{R}_{12}$, we have

$$tr(\mathbf{R}^{-1}\mathbf{B}) = tr(\mathbf{R}^{11}\mathbf{B}_{11} + \mathbf{R}^{12}\mathbf{B}_{21} + \mathbf{R}^{21}\mathbf{B}_{12} + \mathbf{R}^{22}\mathbf{B}_{22})$$

$$= tr((\mathbf{R}_{11}^{-1} + \mathbf{F}\mathbf{R}^{22}\mathbf{F}^{\dagger})\mathbf{B}_{11} - \mathbf{F}\mathbf{R}^{22}\mathbf{B}_{21} - \mathbf{R}^{22}\mathbf{F}^{\dagger}\mathbf{B}_{12} + \mathbf{R}^{22}\mathbf{B}_{22})$$

$$= tr(\mathbf{R}^{22}(\mathbf{B}^{22})^{-1} + \mathbf{R}_{11}^{-1}\mathbf{B}_{11} + (\mathbf{b} - \beta)\mathbf{R}^{22}(\mathbf{b} - \beta)^{\dagger}\mathbf{B}_{11})$$

$$= tr(\mathbf{R}^{22}\mathbf{X} + \mathbf{R}_{11}^{-1}\mathbf{Y} + (\mathbf{Z} - \beta)\mathbf{R}^{22}(\mathbf{Z} - \beta)^{\dagger}\mathbf{Y}). \qquad (3.\mathrm{A}.14)$$

Finally, with (3.A.5), (3.A.12)–(3.A.14) in (3.A.11), the transformed density reduces to

$$f_{\mathbf{X,Y,Z}}(\mathbf{X,Y,Z}) = \frac{|\mathbf{X}|^{N-M_1-M_2}}{\Gamma_{M_2}(N-M_1)\,|(\mathbf{R}^{22})^{-1}|^{N-M_1}}\, e^{-tr(\mathbf{R}^{22}\mathbf{X})}$$

$$\frac{|\mathbf{Y}|^{N-M_1}}{\Gamma_{M_1}(N)\,|(\mathbf{R}_{11})|^{N}}\, e^{-tr(\mathbf{R}_{11}^{-1}\mathbf{Y})}\, \frac{|\mathbf{Y}|^{M_2}}{\pi^{M_1 M_2}\,|(\mathbf{R}^{22})^{-1}|^{M_1}}\, e^{-tr[(\mathbf{Z}-\boldsymbol{\beta})^{\dagger}\mathbf{Y}(\mathbf{Z}-\boldsymbol{\beta})\mathbf{R}^{22}]}$$

$$= f_{\mathbf{X}}(\mathbf{X})f_{\mathbf{Y}}(\mathbf{Y})f_{\mathbf{Z}|\mathbf{Y}}(\mathbf{Z}|\mathbf{Y}). \tag{3.A.15}$$

Notice that both $f_{\mathbf{X}}(\mathbf{X})$ and $f_{\mathbf{Y}}(\mathbf{Y})$ have the form of the complex Wishart distribution and the conditional density $f_{\mathbf{Z}|\mathbf{Y}}(\mathbf{Z}|\mathbf{Y})$ is a quadratic function of $(\mathbf{Z}-\boldsymbol{\beta})$ with a covariance matrix that depends on $\mathbf{Y}$. Further, the Schur component $\mathbf{X} = (\mathbf{B}^{22})^{-1}$ is independent of the regression coefficient $\mathbf{Z} = \mathbf{B}_{11}^{-1}\mathbf{B}_{12}$ and $\mathbf{Y} = \mathbf{B}_{11}$, proving the claim.

Hotelling's Generalized χ^2 Statistic

Let $\mathbf{B} \sim CW(N,M,\mathbf{R})$ and $\mathbf{W}$ an $M \times L$ dimensional matrix of rank L. Then,

$$(\mathbf{W}^{\dagger}\mathbf{B}^{-1}\mathbf{W})^{-1} \sim CW(N-(M-L),L,(\mathbf{W}^{\dagger}\mathbf{R}^{-1}\mathbf{W})^{-1}). \tag{3.A.16}$$

To prove this, define a nonsingular $M \times M$ matrix $\mathbf{X}$ of the form $\mathbf{X} = [\mathbf{Y}\ \mathbf{W}]$ such that the $M \times M-L$ matrix $\mathbf{Y}$ has rank $M-L$ and $\mathbf{Y}^{\dagger}\mathbf{W} = \mathbf{O}$. Then

$$\mathbf{X}^{-1} = \begin{bmatrix} \tilde{\mathbf{Y}}^{-1} \\ \tilde{\mathbf{W}}^{-1} \end{bmatrix}$$

where $\tilde{\mathbf{W}}^{-1}\mathbf{W} = \mathbf{I}_L$, $\tilde{\mathbf{Y}}^{-1}\mathbf{Y} = \mathbf{I}_{M-L}$, $\tilde{\mathbf{W}}^{-1}\mathbf{Y} = \mathbf{O}$ and $\tilde{\mathbf{Y}}^{-1}\mathbf{W} = \mathbf{O}$. In that case

$$\mathbf{S}_w \triangleq \mathbf{X}^{-1}\mathbf{B}(\mathbf{X}^{-1})^{\dagger} = (\mathbf{X}^{\dagger}\mathbf{B}^{-1}\mathbf{X})^{-1}$$

$$= \begin{bmatrix} \mathbf{Y}^{\dagger}\mathbf{B}^{-1}\mathbf{Y} & \mathbf{Y}^{\dagger}\mathbf{B}^{-1}\mathbf{W} \\[2ex] \mathbf{W}^{\dagger}\mathbf{B}^{-1}\mathbf{Y} & \mathbf{W}^{\dagger}\mathbf{B}^{-1}\mathbf{W} \end{bmatrix}^{-1} = (\mathbf{S}_w^{-1})^{-1}. \qquad (3.A.17)$$

Thus, in particular

$$(\mathbf{S}_w^{22})^{-1} = (\mathbf{W}^{\dagger}\mathbf{B}^{-1}\mathbf{W})^{-1}. \qquad (3.A.18)$$

Moreover, $\mathbf{S}_w$ has a complex Wishart distribution. To see this, using (3.A.4), the characteristic function of $\mathbf{S}_w$ simplifies into

$$E\left[e^{j\,tr(\Theta\mathbf{S}_w)}\right] = E\left[e^{j\,tr(\Theta\mathbf{X}^{-1}\mathbf{B}(\mathbf{X}^{-1})^{\dagger})}\right] = E\left[e^{j\,tr([(\mathbf{X}^{-1})^{\dagger}\Theta\mathbf{X}^{-1}]\mathbf{B})}\right]$$

$$= |\,\mathbf{I} - j\,[(\mathbf{X}^{-1})^{\dagger}\Theta\mathbf{X}^{-1}]\mathbf{R}\,|^{-N} = |\,\mathbf{I} - j\,\Theta[\mathbf{X}^{-1}\mathbf{R}(\mathbf{X}^{-1})^{\dagger}]\,|^{-N}$$

or

$$\mathbf{S}_w \sim CW(N,M,\mathbf{X}^{-1}\mathbf{R}(\mathbf{X}^{-1})^{\dagger}).$$

Using (3.A.8), it now follows that the Schur component $(\mathbf{S}_w^{22})^{-1}$ in (3.A.17) has a complex Wishart distribution given by

$$(\mathbf{W}^{\dagger}\mathbf{B}^{-1}\mathbf{W})^{-1} \sim CW(N-(M-L),L,(\mathbf{W}^{\dagger}\mathbf{R}^{-1}\mathbf{W})^{-1}). \quad (3.A.19)$$

In particular, if $\mathbf{W}$ represents an $M \times 1$ vector $\mathbf{w}$, then (3.A.19) together with $c = (\mathbf{w}^{\dagger}\mathbf{R}^{-1}\mathbf{w})^{1/2}$ in (3.A.7) gives

$$\frac{\mathbf{w}^{\dagger}\mathbf{R}^{-1}\mathbf{w}}{\mathbf{w}^{\dagger}\mathbf{B}^{-1}\mathbf{w}} \sim \chi^2(N-M+1). \qquad (3.A.20)$$

Appendix 3.B
Equivalence of Eigenvectors

Let $\tilde{\mathbf{R}}$ be an $M{\times}M$ hermitian matrix with distinct eigenvalues $\tilde{\lambda}_1$, $\tilde{\lambda}_2$, $\cdots$, $\tilde{\lambda}_r$, where m_1, m_2, $\cdots$, m_r, represent their repetitions. Then $m_1 + m_2 + \cdots + m_r = M$; further, let $\tilde{\beta}_{11}$, $\tilde{\beta}_{12}$, $\cdots$, $\tilde{\beta}_{1m_1}$, $\cdots$, $\tilde{\beta}_{1M}$, represent one set of associated normalized eigenvectors. With

$$\tilde{\mathbf{B}}_1 = \left[\tilde{\beta}_{11}, \tilde{\beta}_{12}, \cdots, \tilde{\beta}_{1M} \right] ,$$

$$\tilde{\mathbf{B}}_1 \tilde{\mathbf{B}}_1^\dagger = \mathbf{I}_M$$

and

$$\tilde{\mathbf{\Lambda}} = diag\,[\tilde{\lambda}_1, \tilde{\lambda}_1, \cdots, \tilde{\lambda}_1, \tilde{\lambda}_2, \cdots, \tilde{\lambda}_2, \cdots, \tilde{\lambda}_r\,]$$

we have

$$\tilde{\mathbf{R}} = \tilde{\mathbf{B}}_1 \tilde{\mathbf{\Lambda}} \tilde{\mathbf{B}}_1^\dagger. \tag{3.B.1}$$

Let

$$\tilde{\mathbf{B}}_2 = \left[\tilde{\beta}_{21}, \tilde{\beta}_{22}, \cdots, \tilde{\beta}_{2M} \right]$$

represent yet another set of normalized eigenvectors of $\tilde{\mathbf{R}}$. Then

$$\tilde{\mathbf{R}} = \tilde{\mathbf{B}}_2 \tilde{\mathbf{\Lambda}} \tilde{\mathbf{B}}_2^\dagger, \quad \tilde{\mathbf{B}}_2 \tilde{\mathbf{B}}_2^\dagger = \mathbf{I}_M \tag{3.B.2}$$

and from (3.B.1) and (3.B.2) we have

$$\tilde{\mathbf{B}}_1 \tilde{\mathbf{\Lambda}} \tilde{\mathbf{B}}_1^\dagger = \tilde{\mathbf{B}}_2 \tilde{\mathbf{\Lambda}} \tilde{\mathbf{B}}_2^\dagger$$

or equivalently

$$\tilde{\mathbf{\Lambda}}\mathbf{V} = \mathbf{V}\tilde{\mathbf{\Lambda}} \tag{3.B.3}$$

where

$$\mathbf{V} = \tilde{\mathbf{B}}_1^\dagger \tilde{\mathbf{B}}_2. \tag{3.B.4}$$

Thus, $\mathbf{V}$ is also unitary and, further, $\tilde{\mathbf{\Lambda}}$ and $\mathbf{V}$ commute. Moreover, from (3.B.3) we have

$$\tilde{\lambda}_i v_{ij} = v_{ij} \tilde{\lambda}_j$$

which for $\tilde{\lambda}_i \neq \tilde{\lambda}_j$ gives

$$v_{ij} = 0.$$

This together with the fact that $\mathbf{V}$ is unitary implies $\mathbf{V}$ is block unitary with blocks of sizes $m_1, m_2, \cdots, m_r$; hence from (3.B.4) we have

$$\tilde{\mathbf{B}}_2 = \tilde{\mathbf{B}}_1 \mathbf{V} = \tilde{\mathbf{B}}_1 \begin{bmatrix} \mathbf{V}_1 & & & \mathbf{O} \\ & \mathbf{V}_2 & & \\ & & \ddots & \\ \mathbf{O} & & & \mathbf{V}_r \end{bmatrix} \tag{3.B.5}$$

where

$$\mathbf{V}_i \mathbf{V}_i^\dagger = \mathbf{I}_{m_i}, \quad i = 1, 2, \cdots, r.$$

Notice that in the special case, when all eigenvalues of $\tilde{\mathbf{R}}$ are distinct, then $\mathbf{V}$ is diagonal and unitary and each diagonal entry is a phase factor. In that case

$$\tilde{\beta}_{1i} = e^{j\phi_i} \tilde{\beta}_{2i}, \quad i = 1, 2, \cdots, M. \tag{3.B.6}$$

In particular, (3.B.6) is true for any set of nonrepeating eigenvalues.

Appendix 3.C
Eigenparameters in a Two Source Case

In this appendix, expressions for eigenvalues and eigenvectors of the smoothed covariance matrix $\tilde{\mathbf{R}}$ with $L = 1$ for two coherent signals are derived. Similar results for the conventional MUSIC scheme in an uncorrelated scene are obtained as a special case of this analysis. In addition, several associated inner products that are needed in section 3.3.2 for resolution performance evaluation are also developed.

Consider two coherent sources $\alpha_1 u(t)$ and $\alpha_2 u(t)$ with arrival angles θ_1, θ_2 and source covariance matrix

$$\begin{bmatrix} |\alpha_1|^2 & \alpha_1\alpha_2^* \\ \alpha_2\alpha_1^* & |\alpha_2|^2 \end{bmatrix} = \begin{bmatrix} \alpha_1 \\ \alpha_2 \end{bmatrix} \begin{bmatrix} \alpha_1^* & \alpha_2^* \end{bmatrix} \triangleq \boldsymbol{\alpha}\boldsymbol{\alpha}^\dagger ,$$

$$\alpha_i = |\alpha_i| e^{j\phi_i}, \quad E[\,|u(t)|^2\,] = 1.$$

When the forward/backward smoothing scheme described in section 2.3.3 is deployed once, to decorrelate the coherent signals, from (2.129) the resulting source covariance matrix $\tilde{\mathbf{R}}_u$ has the form

$$\tilde{\mathbf{R}}_u = \frac{1}{2}\left[\boldsymbol{\alpha}\boldsymbol{\alpha}^\dagger + \boldsymbol{\gamma}\boldsymbol{\gamma}^\dagger\right] = \begin{bmatrix} |\alpha_1|^2 & \alpha_1\alpha_2^*\rho_t \\ \alpha_2\alpha_1^*\rho_t^* & |\alpha_2|^2 \end{bmatrix} = \frac{1}{2}[\boldsymbol{\alpha}\ \boldsymbol{\gamma}]\begin{bmatrix}\boldsymbol{\alpha}^\dagger \\ \boldsymbol{\gamma}^\dagger\end{bmatrix} \quad (3.C.1)$$

where

$$\boldsymbol{\gamma} = \begin{bmatrix} \tilde{\nu}_1\alpha_1^* & \tilde{\nu}_2\alpha_2^* \end{bmatrix}^T, \quad \tilde{\nu}_1 = e^{j(M-1)\omega_1}, \quad \tilde{\nu}_2 = e^{j(M-1)\omega_2} \quad (3.C.2)$$

and the effective correlation coefficient ρ_t is given by

$$\rho_t = \frac{1 + e^{-j2(\phi_1 - \phi_2)}\tilde{\nu}_1\tilde{\nu}_2^*}{2} . \quad (3.C.3)$$

Using (3.C.1), the noiseless part of the smoothed covariance matrix $\tilde{\mathbf{R}}$ can be written as

$$\tilde{\mathbf{R}}_1 \triangleq \mathbf{A}\,\tilde{\mathbf{R}}_u\,\mathbf{A}^\dagger = \frac{M}{2}\begin{bmatrix} \mathbf{a}(\omega_1) & \mathbf{a}(\omega_2) \end{bmatrix}\left(\begin{bmatrix} \boldsymbol{\alpha} & \boldsymbol{\gamma} \end{bmatrix}\begin{bmatrix}\boldsymbol{\alpha}^\dagger \\ \boldsymbol{\gamma}^\dagger\end{bmatrix}\right)\begin{bmatrix}\mathbf{a}^\dagger(\omega_1) \\ \mathbf{a}^\dagger(\omega_2)\end{bmatrix}$$

$$\frac{M}{2}\begin{bmatrix} \alpha_1\mathbf{a}(\omega_1) + \alpha_2\mathbf{a}(\omega_2) & \tilde{\nu}_1\alpha_1^*\mathbf{a}(\omega_1) + \tilde{\nu}_2\alpha_2^*\mathbf{a}(\omega_2) \end{bmatrix}\begin{bmatrix} \alpha_1^*\mathbf{a}^\dagger(\omega_1) + \alpha_2^*\mathbf{a}^\dagger(\omega_2) \\ \tilde{\nu}_1^*\alpha_1\mathbf{a}^\dagger(\omega_1) + \tilde{\nu}_2^*\alpha_2\mathbf{a}^\dagger(\omega_2) \end{bmatrix}$$

$$= \frac{1}{2}\begin{bmatrix} \mathbf{b}_1 & \mathbf{b}_2 \end{bmatrix}\begin{bmatrix}\mathbf{b}_1^\dagger \\ \mathbf{b}_2^\dagger\end{bmatrix} \quad (3.C.4)$$

where

$$\mathbf{b}_1 = \sqrt{M}\,(\alpha_1 \mathbf{a}(\omega_1) + \alpha_2 \mathbf{a}(\omega_2)) \tag{3.C.5}$$

and

$$\mathbf{b}_2 = \sqrt{M}\,(\tilde{\nu}_1 \alpha_1^* \mathbf{a}(\omega_1) + \tilde{\nu}_2 \alpha_2^* \mathbf{a}(\omega_2))\,. \tag{3.C.6}$$

The nonzero eigenvalues of $\tilde{\mathbf{R}}_1$ are given by the roots of the quadratic equation

$$\tilde{\mu}^2 - tr(\mathbf{A}\tilde{\mathbf{R}}_u \mathbf{A}^\dagger)\tilde{\mu} + |\tilde{\mathbf{R}}_u \mathbf{A}^\dagger \mathbf{A}| = 0\,, \tag{3.C.7}$$

and for $i = 1, 2$, they are

$$\tilde{\mu}_i = \frac{1}{2} tr(\mathbf{A}\tilde{\mathbf{R}}_u \mathbf{A}^\dagger)\left[1 \pm \left(1 - \frac{|\tilde{\mathbf{R}}_u \mathbf{A}^\dagger \mathbf{A}|}{\left[tr(\mathbf{A}\tilde{\mathbf{R}}_u \mathbf{A}^\dagger)/2\right]^2}\right)^{1/2}\right]\,. \tag{3.C.8}$$

To evaluate $\tilde{\mu}_1$, $\tilde{\mu}_2$, we define the spatial correlation coefficient ρ_s between the sources for an M element array to be

$$\rho_s = \mathbf{a}^\dagger(\omega_1)\,\mathbf{a}(\omega_2) = \frac{1}{M} e^{j(M-1)\omega_d}\,\frac{\sin M\omega_d}{\sin \omega_d}$$

$$= e^{j(M-1)\omega_d} Si(M\omega_d) \tag{3.C.9}$$

with

$$\omega_d = (\omega_1 - \omega_2)/2\,. \tag{3.C.10}$$

Here, by definition

$$Si(M\omega_d) = \frac{1}{M}\,\frac{\sin M\omega_d}{\sin \omega_d}\,. \tag{3.C.11}$$

From (3.C.4) – (3.C.6) we have

$$tr(\mathbf{A}\tilde{\mathbf{R}}_u \mathbf{A}^\dagger) = \frac{1}{2}(\mathbf{b}_1^\dagger \mathbf{b}_1 + \mathbf{b}_2^\dagger \mathbf{b}_2)$$

$$= M\,(\,|\alpha_1|^2 + |\alpha_2|^2 + Re\,(\alpha_1^* \alpha_2 \rho_s + \alpha_1 \alpha_2 \tilde{\nu}_1^* \tilde{\nu}_2 \rho_s)) \tag{3.C.12}$$

and

$$| \tilde{\mathbf{R}}_u \mathbf{A}^\dagger \mathbf{A} | \; = \; | \tilde{\mathbf{R}}_u | \; | \mathbf{A}^\dagger \mathbf{A} |$$

$$= M^2 \, |\alpha_1|^2 \, |\alpha_2|^2 (1 - |\rho_t|^2)(1 - |\rho_s|^2) \,. \quad (3.C.13)$$

From here onwards we will consider the two sources to be perfectly coherent and of equal power; i.e., $\alpha_1 = \alpha_2$ and $|\alpha_1|^2 = 1$. In that case

$$\rho_t \; = \; \frac{1 + \tilde{\nu}_1 \tilde{\nu}_2^*}{2} \; = \; e^{\,j(M-1)\omega_d} \cos\big((M-1)\omega_d\big) \quad (3.C.14)$$

$$\mathrm{Re}(\rho_s) = \cos[(M-1)\omega_d]\, Si(M\omega_d) = \mathrm{Re}(\rho_s \rho_t^*) \quad (3.C.15)$$

and with (3.C.12) – (3.C.13) in (3.C.8) it is easy to see that

$$\tilde{\mu}_i = M\left[1 + \mathrm{Re}(\rho_s \rho_t^*) \pm \left[\Big(1 + \mathrm{Re}(\rho_s \rho_t^*)\Big)^2 \right.\right.$$

$$\left.\left. - \Big(1 - |\rho_s|^2\Big)\Big(1 - |\rho_t|^2\Big)\right]^{1/2}\right]$$

$$= M\left[1 + \mathrm{Re}(\rho_s \rho_t^*) \pm |\rho_s + \rho_t|\right], \quad i = 1, 2, \quad (3.C.16)$$

where we have made use of the identity $[\mathrm{Re}(\rho_s \rho_t^*)]^2 = |\rho_s|^2 |\rho_t|^2$. The eigenvectors corresponding to these eigenvalues span the two dimensional signal subspace spanned by $\mathbf{a}(\omega_1)$ and $\mathbf{a}(\omega_2)$, and from (3.C.4) they are linear combinations of $\mathbf{b}_1$ and $\mathbf{b}_2$; i.e.,

$$\tilde{\beta}_i \propto (\mathbf{b}_1 + k_i \, \mathbf{b}_2), \quad i = 1, 2. \quad (3.C.17)$$

Moreover, $\tilde{\mu}_i, \tilde{\beta}_i, i = 1, 2$, as a pair satisfy

$$(\mathbf{A}\tilde{\mathbf{R}}_u \mathbf{A}^\dagger)\tilde{\beta}_i = \tilde{\mu}_i \tilde{\beta}_i \,, \quad i = 1, 2,$$

which together with (3.C.4) results in

$$\frac{1}{2}[\mathbf{b}_1^\dagger\mathbf{b}_1 + k_i\,\mathbf{b}_1^\dagger\mathbf{b}_2]\,\mathbf{b}_1 + \frac{1}{2}[\mathbf{b}_2^\dagger\mathbf{b}_1 + k_i\,\mathbf{b}_2^\dagger\mathbf{b}_2]\,\mathbf{b}_2 = \bar{\mu}_i\,\mathbf{b}_1 + \bar{\mu}_i k_i\,\mathbf{b}_2\,,$$

$$i = 1, 2. \qquad (3.C.18)$$

The solution to (3.C.17) need not be unique. However, since the eigenvectors can be made unique by proper normalization, at this stage we seek a solution set to the distinct equations

$$\frac{1}{2}[\mathbf{b}_1^\dagger\mathbf{b}_1 + k_i\,\mathbf{b}_1^\dagger\mathbf{b}_2] = \bar{\mu}_i \qquad (3.C.19)$$

and

$$\frac{1}{2}[\mathbf{b}_2^\dagger\mathbf{b}_1 + k_i\,\mathbf{b}_2^\dagger\mathbf{b}_2] = \bar{\mu}_i k_i\,, \quad i = 1, 2. \qquad (3.C.20)$$

Solutions to (3.C.19) − (3.C.20), if they exist, satisfy (3.C.18). Further, the consistency of the above equations can be verified by observing that the solution

$$k_i = (2\bar{\mu}_i - \mathbf{b}_1^\dagger\mathbf{b}_1)/\mathbf{b}_1^\dagger\mathbf{b}_2 \qquad (3.C.21)$$

from (3.C.19), when substituted into (3.C.20), results in

$$\frac{4}{\mathbf{b}_1^\dagger\mathbf{b}_2}\left[\bar{\mu}_i^2 - \frac{1}{2}(\mathbf{b}_1^\dagger\mathbf{b}_1 + \mathbf{b}_2^\dagger\mathbf{b}_2)\bar{\mu}_i + \frac{1}{4}\Big((\mathbf{b}_1^\dagger\mathbf{b}_1)(\mathbf{b}_2^\dagger\mathbf{b}_2) - |\,\mathbf{b}_1^\dagger\mathbf{b}_2\,|^2\Big)\right] = 0.$$

Using (3.C.12) − (3.C.13), since $\mathbf{b}_1^\dagger\mathbf{b}_2 \neq 0$, it is easy to verify that this equation is the same as (3.C.7), proving our claim. To simplify k_i ; $i = 1, 2$, further, using (3.C.15), notice that

$$\mathbf{b}_1^\dagger\mathbf{b}_1 = M\,(\,|\alpha_1|^2 + |\alpha_2|^2 + 2\,\mathrm{Re}(\alpha_1\alpha_2^*\rho_s))$$

$$= 2M\,(1 + Re\,(\rho_s)) = 2M\,(1 + \mathrm{Re}(\rho_s\rho_t^*))$$

$$\mathbf{b}_1^\dagger\mathbf{b}_2 = \alpha_1^{*2}M\,[\bar{\nu}_1(1 + \rho_s^*) + \bar{\nu}_2(1 + \rho_s)]$$

$$= 2\alpha_1^{*2}M\,\bar{\nu}_1(\rho_s^* + \rho_t^*) = 2\alpha_1^{*2}M\,\bar{\nu}_2(\rho_s + \rho_t)$$

and these together with (3.C.16) in (3.C.21) yield

$$k_i = \pm \frac{|\,\rho_s + \rho_t\,|}{\alpha_1^{*^2}\,\tilde{\nu}_1(\rho_s^* + \rho_t^*)} = \pm \frac{|\,\rho_s + \rho_t\,|}{\alpha_1^{*^2}\,\tilde{\nu}_2(\rho_s + \rho_t)}. \qquad (3.C.22)$$

Using (3.C.22), the eigenvectors in (3.C.17) can be written as

$$\tilde{\beta}_i \propto (\mathbf{b}_1 + k_i \mathbf{b}_2) = (\alpha_1 + \alpha_1^* \tilde{\nu}_1 k_i)\mathbf{a}(\omega_1) + (\alpha_1 + \alpha_1^* \tilde{\nu}_2 k_i)\mathbf{a}(\omega_2)$$

$$\propto \left[1 \pm \frac{|\,\rho_s + \rho_t\,|}{\rho_s^* + \rho_t^*}\right]\mathbf{a}(\omega_1) + \left[1 \pm \frac{|\,\rho_s + \rho_t\,|}{\rho_s + \rho_t}\right]\mathbf{a}(\omega_2). \qquad (3.C.23)$$

To simplify this further, observe that

$$\rho_s + \rho_t = e^{j(M-1)\omega_d}\Big[\cos\big((M-1)\omega_d\big) + Si(M\omega_d)\Big]$$

and consider the case when $[\cos((M-1)\omega_d) + Si(M\omega_d)] > 0$. Then from (3.C.23)

$$\tilde{\beta}_i \propto e^{j(M-1)\omega_1/2}\left[e^{-j(M-1)\omega_1/2} \pm e^{-j(M-1)\omega_2/2}\right]\mathbf{a}(\omega_1)$$

$$\pm e^{j(M-1)\omega_2/2}\left[e^{-j(M-1)\omega_1/2} \pm e^{-j(M-1)\omega_2/2}\right]\mathbf{a}(\omega_2)$$

$$\propto \mathbf{u}_1 \pm \mathbf{u}_2$$

where

$$\mathbf{u}_1 = e^{j(M-1)\omega_1/2}\mathbf{a}(\omega_1), \qquad \mathbf{u}_2 = e^{j(M-1)\omega_2/2}\mathbf{a}(\omega_2). \qquad (3.C.24)$$

From the above discussion, it now follows that

$$\tilde{\beta}_i \propto \begin{cases} \mathbf{u}_1 \pm \mathbf{u}_2 & [\cos((M-1)\omega_d) + Si(M\omega_d)] > 0 \\ \mathbf{u}_1 \mp \mathbf{u}_2 & \text{otherwise} \end{cases}$$

$$i = 1, 2$$

Finally, the normalized eigenvectors corresponding to the nonzero eigenvalues in (3.C.16) are given by

$$
\tilde{\beta}_i = \begin{cases}
(\mathbf{u}_1 \pm \mathbf{u}_2)/\sqrt{2(1 \pm Si(M\omega_d))} & [\cos((M-1)\omega_d) + Si(M\omega_d)] > 0 \\[2ex]
(\mathbf{u}_1 \mp \mathbf{u}_2)/\sqrt{2(1 \mp Si(M\omega_d))} & \text{otherwise}
\end{cases}
$$

$$
i = 1, 2. \qquad (3.C.25)
$$

With $\mathbf{u}_1$, $\mathbf{u}_2$ as given in (3.C.24), the eigenvalues and eigenvectors for an uncorrelated source scene with equipowered signals drop out as a special case of this analysis. In that situation, the source covariance matrix $\mathbf{R}_u$ has the form

$$
\begin{bmatrix} 1 & 0 \\ 0 & 1 \end{bmatrix}.
$$

Notice that this is a special case of (3.C.1) with $\alpha_1 = \alpha_2 = 1$, and $\rho_t = 0$. Thus from (3.C.16)

$$
\mu_i = M(1 \pm |\rho_s|) \qquad (3.C.26)
$$

and similarly

$$
\beta_i = \begin{cases}
(\mathbf{u}_1 \pm \mathbf{u}_2)/\sqrt{2(1 \pm Si(M\omega_d))} & Si(M\omega_d) > 0 \\[2ex]
(\mathbf{u}_1 \mp \mathbf{u}_2)/\sqrt{2(1 \mp Si(M\omega_d))} & \text{otherwise}
\end{cases}
$$

$$
i = 1, 2. \qquad (3.C.27)
$$

From (3.C.25) and (3.C.27) we can conclude that for equipowered sources, irrespective of their effective correlation ρ_t resulting from spatial smoothing, the smoothed and the uncorrelated cases have the same set of normalized eigenvectors whenever $[\cos((M-1)\omega_d) + Si(M\omega_d)]$ and $Si(M\omega_d)$ have the same sign; i.e.,

$$\tilde{\beta}_i = \beta_i , \quad i = 1, 2 \qquad [\cos((M-1)\omega_d) + Si(M\omega_d)] \gtrless 0$$

$$\text{and } Si(M\omega_d) \gtrless 0$$

$$\tilde{\beta}_1 = \beta_2 , \quad \tilde{\beta}_2 = \beta_1 \qquad \text{otherwise}.$$

We conclude this appendix with several useful parametric approximations to eigenvalues, and inner products between eigenvectors and direction vectors for both uncorrelated and coherent cases. To start with, let [10]

$$\Delta^2 \triangleq \frac{M^2 \omega_d^2}{3} . \tag{3.C.28}$$

For closely spaced sources $(M\omega_d)^2 \ll 1$, and in that case from (3.C.14) and (3.C.9)

$$\rho_t = e^{-j(M-1)\omega_d} \left[1 - \frac{1}{2}(M-1)^2 \omega_d^2 + \frac{1}{24}(M-1)^4 \omega_d^4 + \cdots \right]$$

$$= e^{-j(M-1)\omega_d} \left[1 - \frac{3}{2}\left(\frac{M-1}{M}\right)^2 \Delta^2 + \frac{3}{8}\left(\frac{M-1}{M}\right)^4 \Delta^4 + \cdots \right] \tag{3.C.29}$$

and similarly

$$Si(M\omega_d) = \frac{1}{M}\frac{\sin M\omega_d}{\sin \omega_d} \simeq \left[1 - \frac{1}{2}\Delta^2 + \frac{3}{40}\Delta^4 \right] \tag{3.C.30}$$

$$\rho_s \simeq e^{j(M-1)\omega_d} \left[1 - \frac{1}{2}\Delta^2 + \frac{3}{40}\Delta^4 \right] . \tag{3.C.31}$$

For the uncorrelated case parametric expressions for the eigenvalues can be easily obtained from (3.C.26) and (3.C.31). In a similar manner for the coherent case, using (3.C.16) we get

$$\tilde{\mu}_1 \simeq 4M \left[1 - \left(1 - \frac{3}{2M} \right) \Delta^2 + \left(\frac{33}{80} - \frac{9}{8M} \right) \Delta^4 \right] \qquad (3.C.32)$$

$$\tilde{\mu}_2 \simeq \frac{3}{4} M \left(1 - \frac{2}{M} \right) \Delta^4 . \qquad (3.C.33)$$

Because of the equivalence of eigenvectors for two uncorrelated and perfectly coherent equipowered sources, we have (for $Si(M\omega_d) > 0$)

$$| \tilde{\beta}_i^{\dagger} \mathbf{a}(\omega) |^2 = | \beta_i^{\dagger} \mathbf{a}(\omega) |^2 , \quad \text{for all } \omega, \quad i = 1, 2 \qquad (3.C.34)$$

and it suffices to obtain the inner products of the eigenvectors in (3.C.27) with direction vectors associated with the true arrival angles. A little algebra shows that

$$| \tilde{\beta}_1^{\dagger} \mathbf{a}(\omega_1) |^2 = | \beta_1^{\dagger} \mathbf{a}(\omega_2) |^2 = \frac{1 + Si(M\omega_d)}{2}$$

$$\simeq 1 - \frac{1}{4} \Delta^2 + \frac{3}{80} \Delta^4 \qquad (3.C.35)$$

$$| \tilde{\beta}_2^{\dagger} \mathbf{a}(\omega_1) |^2 = | \beta_2^{\dagger} \mathbf{a}(\omega_2) |^2 = \frac{1 - Si(M\omega_d)}{2}$$

$$\simeq \frac{1}{4} \Delta^2 - \frac{3}{80} \Delta^4 . \qquad (3.C.36)$$

Similarly, for the mid-arrival angle

$$\omega_m = (\omega_1 + \omega_2)/2$$

we also have

$$| \tilde{\beta}_1^{\dagger} \mathbf{a}(\omega_m) |^2 = | \beta_1^{\dagger} \mathbf{a}(\omega_m) |^2 = \frac{2 (Si(M\omega_d/2))^2}{1 + Si(M\omega_d)}$$

$$\simeq 1 - \frac{1}{80} \Delta^4 \qquad (3.C.37)$$

and

$$| \tilde{\beta}_2^\dagger \mathbf{a}(\omega_m) |^2 \ = \ | \beta_2^\dagger \mathbf{a}(\omega_m) |^2 \simeq 0. \qquad (3.C.38)$$

Finally, with $\xi = \tilde{\xi} \triangleq PM/\sigma^2 = M/\sigma^2$ representing the output SNR, and using (3.C.32) $-$ (3.C.36) in (3.81) $-$ (3.82), for bias we have

$$\eta(\omega_i) \ = \ \frac{M-2}{N} \left[\frac{1}{\xi} + \frac{\Delta^{-2}}{\xi^2} \right], \quad i = 1, 2 \qquad (3.C.39)$$

$$\tilde{\eta}(\omega_i) \ = \ \frac{M-2}{N} \left[\frac{1}{\tilde{\xi}} \left(\frac{M\Delta^{-2}}{6(M-2)} + \frac{1}{10} \right) \right.$$

$$\left. + \frac{\Delta^{-2}}{\tilde{\xi}^2} \left(\frac{2M\Delta^{-4}}{9(M-4)} - \frac{M\Delta^{-2}}{30(M-4)} \right) \right] ;$$

$$i = 1, 2 \qquad (3.C.40)$$

and similarly for the mid-arrival angle ω_m, this gives

$$E[\hat{Q}(\omega_m)] = a + \frac{1}{N} \left[\frac{1}{\xi} b + \frac{1}{\xi^2} c \right] \qquad (3.C.41)$$

in an uncorrelated source scene and

$$E[\hat{Q}(\omega_m)] = \tilde{a} + \frac{1}{N} \left[\frac{1}{\tilde{\xi}} \tilde{b} + \frac{1}{\tilde{\xi}^2} \tilde{c} \right] \qquad (3.C.42)$$

in a coherent scene. Here

$$a = \tilde{a} = Q(\omega_m) \simeq \Delta^4/80 ,$$

$$b \simeq \left(\frac{M-2}{2} (1 + \Delta^2/4) \right), \quad c \simeq \left(\frac{M-2}{4} (1 + \Delta^2/2) \right),$$

$$\tilde{b} \simeq \frac{M-2}{4} \left[1 + \left(1 - \frac{3}{2M} \right) \Delta^2 \right]$$

and

$$\bar{c} \simeq \frac{M-2}{16}\left[1 + \left(2 - \frac{3}{M}\right)\Delta^2\right] - \frac{M\Delta^{-4}}{45(M-4)} \, .$$

As a consequence of the above analysis, we also have

$$\sigma^2(\omega_m) = \frac{d}{N}\left[\frac{1}{2\xi}e + \frac{1}{4\xi^2}f\right] \tag{3.C.43}$$

and

$$\tilde{\sigma}^2(\omega_m) = \frac{\tilde{d}}{N}\left[\frac{1}{4\xi}\tilde{e} + \frac{1}{16\xi^2}\tilde{f}\right] \tag{3.C.44}$$

where

$$d = \tilde{d} = \frac{\Delta^4}{40}\left[1 - \frac{\Delta^4}{80}\right]$$

$$e \simeq 1 + \Delta^2/4 + \Delta^4/40, \quad f \simeq 1 + \Delta^2/2 + \Delta^4/80,$$

$$\tilde{e} \simeq 1 + (1 - 3/2M)\Delta^2 + (47/80 - 15/8M + 9/4M^2)\Delta^4,$$

$$\tilde{f} \simeq 1 + (2 - 3/M)\Delta^2 + (87/40 - 27/4M + 27/4M^2)\Delta^4.$$

Problems

1. **The Cramer-Rao bound.** Let $\theta = [\theta_1, \theta_2, \cdots, \theta_K]^T$ represent the unknown (nonrandom) parameters present in the data vectors $\mathbf{x}(n)$, $n = 1, 2, \cdots, N$ given by (3.1). Further let $\hat{\theta} = [\hat{\theta}_1(\mathbf{x}), \hat{\theta}_2(\mathbf{x}), \cdots, \hat{\theta}_K(\mathbf{x})]^T$ represent any unbiased estimator for the unknown parameter set θ. In general the covariance matrix

$$Cov(\hat{\theta}) = E[(\hat{\theta}(\mathbf{x}) - \theta)(\hat{\theta}(\mathbf{x}) - \theta)^\dagger] \tag{3.P.1}$$

will depend on the actual form of the estimate $\hat{\theta}(\mathbf{x})$. However, it is possible to obtain the minimum of (3.P.1) over the class of all possible estimators without actually knowing the corresponding estimate. This remarkable result, known as the Cramer-Rao bound, states that

$$Cov(\hat{\boldsymbol{\theta}}) \geq \mathbf{J}^{-1}(\boldsymbol{\theta})$$

where the inequality applies term by term. Here $\mathbf{J}(\boldsymbol{\theta})$ is the Fisher information matrix, whose $(i,j)^{\text{th}}$ element is given by

$$J_{ij} = E\left[\frac{\partial \ln L(\mathbf{x}(1), \cdots, \mathbf{x}(n) \mid \boldsymbol{\theta})}{\partial \theta_i} \frac{\partial \ln L(\mathbf{x}(1), \cdots, \mathbf{x}(n) \mid \boldsymbol{\theta})}{\partial \theta_j} \right]$$

$$= -E\left[\frac{\partial^2 \ln L(\mathbf{x}(1), \mathbf{x}(2), \cdots, \mathbf{x}(n) \mid \boldsymbol{\theta})}{\partial \theta_i \, \partial \theta_j} \right].$$

Here $\ln L(\mathbf{x}(1), \mathbf{x}(2), \cdots, \mathbf{x}(n) \mid \boldsymbol{\theta})$ is the log-likelihood function of $\mathbf{x}(1), \mathbf{x}(2), \cdots, \mathbf{x}(n)$ given $\boldsymbol{\theta}$.

a) Using (3.3) – (3.5) show that

$$\frac{\partial \ln L(\mathbf{x}(1), \mathbf{x}(2), \cdots, \mathbf{x}(n) \mid \boldsymbol{\theta})}{\partial \theta_i} = N \, tr\left[\frac{\partial \mathbf{R}^{-1}}{\partial \theta_i} (\mathbf{R} - \hat{\mathbf{R}}) \right]$$

and hence

$$J_{ij} = -N \, tr\left(\frac{\partial \mathbf{R}^{-1}}{\partial \theta_i} \frac{\partial \mathbf{R}}{\partial \theta_j} \right) = N \, tr\left(\mathbf{R}^{-1} \frac{\partial \mathbf{R}}{\partial \theta_i} \mathbf{R}^{-1} \frac{\partial \mathbf{R}}{\partial \theta_j} \right).$$

b) Show that the bound on the variance of any unbiased estimate $\hat{\theta}_k(\mathbf{x})$ for the unknown parameter θ_k has the form

$$Var(\hat{\theta}_k) \geq J^{kk} = \left(\mathbf{J}^{-1} \right)_{kk}. \tag{3.P.2}$$

c) When θ_k is the only unknown parameter (i.e., $\theta_1, \cdots, \theta_{k-1}, \theta_{k+1}, \cdots, \theta_K$ are known), show that the corresponding bound is

$$Var(\hat{\theta}_k \mid \theta_1, \cdots, \theta_{k-1}, \theta_{k+1}, \cdots, \theta_K) \geq 1/J_{kk}. \tag{3.P.3}$$

Conclude that the bound in (3.P.2) is always inferior to that in (3.P.3).

2.   **Symmetric array scheme.** Let $\hat{Q}_s(\omega)$ represent the sample estimator associated with $Q_s(\omega)$ in (2.100).

  a)   Show that $\hat{Q}_s(\omega)$ is a consistent estimator.

  b)   Within a first order approximation, show that the null estimator possesses zero variance along the true directions of arrival, i.e., $Var[\hat{Q}_s(\omega_k)] = 0, k = 1, 2, \cdots$.

  c)   **Two source scene.** With symbols as in text show that

$$E[\hat{Q}_s(\omega_k)] = \frac{\mu(1+2\mu\xi)}{2N(\mu\xi)^2}\left[\frac{1}{30} + \frac{16}{35}\frac{1}{\Delta^2}\right], \quad k = 1, 2$$

and

$$E[\hat{Q}_s(\omega_m)] = \frac{\Delta^4}{80} + \frac{\mu(1+2\mu\xi)}{2N(\mu\xi)^2}\left[\frac{1}{30} + \frac{16}{175}\frac{1}{\Delta^2}\right]$$

where $\xi = MP/\sigma^2$ and $\mu = 1+Re(\rho_{12})$. Notice that $\mu = 0$ is excluded by the requirement $b_0 \neq 0$ in (2.82). From (3.84), this gives the resolution threshold to be

$$\xi_s = \frac{512}{35\Delta^6 N\mu}\left[1 + \left[1 + \frac{35}{512}\Delta^6 MN\right]^{1/2}\right].$$

3.   **Performance analysis for the mean based scheme** [30]. For data vectors $\mathbf{x}(t_n)$, $n = 1, 2, \cdots, N$ with nonzero mean signals, define with

$$\hat{c}_i = \frac{1}{N}\sum_{n=1}^{N}\left[\frac{x_i(t_n) + x_{-i}^*(t_n)}{2}\right], \quad i = 0, 1, \cdots, (M-1).$$

Let $\hat{T}_2$ represent the hermitian Toeplitz matrix generated from these sample mean values and

$$\hat{D}(\omega) = \sum_{i=K+1}^{M} |\hat{u}_i^\dagger \mathbf{a}(\omega)|^2$$

the associated sample null estimator. Here $\hat{u}_i$, $i = K+1, K+2, \cdots, M$ are the sample eigenvectors associated with the estimates of the zero eigenvalue in $\hat{\mathbf{T}}_2$.

(a)   Using the results in section 3.3.1 show that

$$E[\hat{D}(\omega)] = D(\omega)$$

$$- \frac{1}{N} \sum_{\substack{i=1 \\ }}^{K} \sum_{\substack{k=1 \\ k \neq i}}^{M} \sum_{\substack{l=1 \\ l \neq i}}^{M} \frac{\mathbf{u}_k^\dagger \mathbf{B}(i,i) \mathbf{u}_l}{(\lambda_i - \lambda_k)(\lambda_i - \lambda_l)} \mathbf{a}^\dagger(\omega) \mathbf{u}_k \, \mathbf{u}_l^\dagger \mathbf{a}(\omega)$$

$$+ \frac{1}{N} \sum_{\substack{i=1 \\ }}^{K} \sum_{\substack{j=1 \\ j \neq i}}^{M} \frac{\mathbf{u}_j^\dagger \mathbf{B}(i,i) \mathbf{u}_j}{(\lambda_i - \lambda_j)^2} \, | \mathbf{u}_i^\dagger \mathbf{a}(\omega) |^2 + o(\frac{1}{N}) \quad (3.\text{P}.4)$$

(b)   and

$$Var[\hat{D}(\omega)]$$

$$= \frac{2}{N} \sum_{\substack{i=1 \\ }}^{K} \sum_{\substack{j=1 \\ }}^{K} \sum_{\substack{k=1 \\ k \neq i}}^{M} \sum_{\substack{l=1 \\ l \neq j}}^{M} \frac{\text{Re}\left[\mathbf{a}^\dagger(\omega) \mathbf{u}_k \, \mathbf{u}_i^\dagger \mathbf{a}(\omega) \Delta_{ijkl} \right]}{(\lambda_i - \lambda_k)(\lambda_j - \lambda_l)}$$

$$+ o(\frac{1}{N}) \qquad\qquad (3.\text{P}.5)$$

where

$$\Delta_{ijkl} \triangleq \mathbf{u}_k^\dagger \mathbf{B}(i,l) \mathbf{u}_j \, \mathbf{a}^\dagger(\omega) \mathbf{u}_l \, \mathbf{u}_j^\dagger \mathbf{a}(\omega)$$

$$+ \mathbf{u}_k^\dagger \mathbf{B}(i,j) \mathbf{u}_l \, \mathbf{a}^\dagger(\omega) \mathbf{u}_j \, \mathbf{u}_i^\dagger \mathbf{a}(\omega)$$

$$= \Delta_{ilkj}$$

and the matrix $\mathbf{B}(i,j)$ is independent of N and is given by

$$[\mathbf{B}(i,j)]_{kl} = \mathbf{u}_j^\dagger \left[M\,(\mathbf{A}_l \mathbf{A}^* + \mathbf{B}_l \mathbf{A}) \mathbf{R}_s (\mathbf{A}_k \mathbf{A}^* + \mathbf{B}_k \mathbf{A})^\dagger \right.$$

$$\left. + \mathbf{A}_l \mathbf{R}_w \mathbf{A}_k^\dagger + \mathbf{B}_l \mathbf{R}_w \mathbf{B}_k^\dagger \right] \mathbf{u}_i$$

with

$$A_1 = \begin{bmatrix} 1 & & \\ & & \\ & & O \end{bmatrix}, \quad B_1 = \begin{bmatrix} 0 & & & \\ & 1 & & \\ & & 1 & \\ & O & & \ddots & \\ & & & & 1 \end{bmatrix}$$

$$A_i = \begin{bmatrix} O & & 1 \\ & \ddots & \\ 1 & & \\ & & O \end{bmatrix}, \quad B_i = \begin{bmatrix} & & & O \\ 0 & & & \\ & 1 & & \\ & & \ddots & \\ O & & & 1 \end{bmatrix},$$

$$i = 2, 3, \cdots, M\text{-}1,$$

$$A_M = \begin{bmatrix} O & & & 1 \\ & & 1 & \\ & \ddots & & \\ & 1 & & O \\ 1 & & & \end{bmatrix}, \quad B_M = O.$$

Further,

$$\mathbf{R}_s \triangleq E[(\mathbf{u}_r(t)-\mathbf{b})(\mathbf{u}_r(t)-\mathbf{b})^\dagger],$$

$$\mathbf{u}_r(t) = [\operatorname{Re}(u_1(t)), \operatorname{Re}(u_2(t)), \cdots, \operatorname{Re}(u_K(t))]^T$$

and

$$\mathbf{R}_w = E[\mathbf{w}(t)\mathbf{w}^\dagger(t)],$$

$$\mathbf{w}(t) = [w_0(t), w_1(t), \cdots, w_{M\text{-}1}(t)]^T$$

with

$$w_i(t) = \frac{n_i(t) + n^*_{-i}(t)}{2}, \quad i = 0, 1, \cdots, M-1.$$

(c) Using (3.P.5) show that, within a $1/N$ approximation,

$$Var[\hat{D}(\omega_k)] = 0, \quad k = 1, 2, \cdots, K.$$

(d) If $\mathbf{u}_i$ or $\mathbf{u}_j$ in (3.P.5) is a noise eigenvector, for uncorrelated noise of equal variance σ^2 show that

$$\mathbf{u}_k^\dagger \mathbf{B}(i, j) \mathbf{u}_l$$

$$= \frac{\sigma^2}{2} \mathbf{u}_i^\dagger
\begin{bmatrix}
\delta_{kl} & \sum_{r=1}^{M-1} u^*_{r+1,k} u_{rl} & \cdots & u^*_{Mk} u_{1l} \\[2ex]
\sum_{r=1}^{M-1} u^*_{rk} u_{r+1,l} & \delta_{kl} & \cdots & \sum_{r=1}^{2} u^*_{r+M-2,k} u_{rl} \\[2ex]
\vdots & \vdots & \ddots & \vdots \\[2ex]
u^*_{1k} u_{Ml} & \sum_{r=1}^{2} u^*_{rk} u_{r+M-2,l} & \cdots & \delta_{kl}
\end{bmatrix}
\mathbf{u}_j$$

$$\triangleq \frac{\sigma^2}{2} \mathbf{u}_i^\dagger \mathbf{E}(k, l) \mathbf{u}_j = \frac{\sigma^2}{2} \mathbf{u}_k^\dagger \mathbf{E}(i, j) \mathbf{u}_l. \tag{3.P.6}$$

(e) **Single source case.** With

$$\beta = \frac{|\text{ signal mean }|^2}{\text{total signal power}} = \frac{2b^2}{P}$$

under the assumption that $b = E[\text{Re}(u_1(t))] = E[\text{Im}(u_1(t))]$, using (3.P.6) show that

$$E[\hat{D}(\omega_1)] = \frac{\sigma^2}{N M \beta P}\left[\frac{1}{3} - \frac{1}{3M^2}\right] + o\left(\frac{1}{N}\right)$$

$$\simeq \frac{\sigma^2}{3N M \beta P}\,.$$

(f) **Two source case.** By making use of appropriate results developed in Appendix C show that

$$E[\hat{D}(\omega_1)] \simeq \frac{\sigma^2}{2N M b^2}\left[\frac{1}{30} + \frac{16}{35}\frac{1}{\Delta^2}\right],$$

$$E[\hat{D}(\omega_m)] \simeq \frac{\Delta^4}{80} + \frac{\sigma^2}{2N M b^2}\left[\frac{1}{30} + \frac{16}{175}\frac{1}{\Delta^2}\right]$$

where $b_1 = b_2 = b$ and finally the resolution threshold to be

$$\xi_m \simeq \frac{29}{N \beta \Delta^6}\,. \tag{3.P.7}$$

4. If all eigenvalues of $\mathbf{R}$ in (3.2) are distinct, then show that $tr(\mathbf{B})$ is a sum of M independent Gamma-distributed random variables. Find the parameters of these Gamma random variables. (Hint: Let $\Theta = \theta\mathbf{I}$ in (3.A.4).)

5. For any $M \times 1$ vector $\mathbf{w}$ show that [31 - 33]

$$z = \frac{(\mathbf{w}^\dagger \mathbf{B}^{-1}\mathbf{w})^2}{(\mathbf{w}^\dagger \mathbf{B}^{-1}\mathbf{R}\mathbf{B}^{-1}\mathbf{w})}\frac{1}{(\mathbf{w}^\dagger \mathbf{R}^{-1}\mathbf{w})} \sim \beta(N-M+2, M-1) \tag{3.P.8}$$

i.e., $\mathbf{z}$ is a Beta-distributed random variable with parameters $N-M+2$ and $M-1$. Here $\mathbf{R}$ and $\mathbf{B}$ are given in (3.2) and (3.4).

6. Consider the two source scenario described in section 3.3.2.

a) Show that in the case of uncorrelated sources, when the f/b scheme is deployed once ($L = 1$), the new resolution threshold can be expressed as

$$\tilde{\xi}_{T,\mu} \simeq \frac{1}{N}\left[\frac{10\,(M-2)}{\Delta^4}\left(1+\left(1+\frac{2N}{5\,(M-2)}\Delta^2\right)^{1/2}\right)\right] = \xi_T/2.$$

Here ξ_T is as given in (3.85).

b)   Complete the performance analysis evaluation for forward-only smoothing scheme that makes use of two overlapping subarrays and compare the new resolution threshold with (3.85) and (3.86).

7.   **GEESE scheme: Uncorrelated case** [34].

(a)   Show that in a two uncorrelated source scene, the least favorable configuration for the GEESE scheme gives rise to unbiased estimators $\hat{\gamma}_1$ and $\hat{\gamma}_2$ (derive (3.92)). Evaluate the variance associated with these estimators.

(b)   In its most favorable configuration, show that the GEESE scheme gives rise to biased estimators in a two source scene. Obtain expressions for bias and variance associated with these estimators.

8.   **GEESE scheme: Coherent case** A two coherent source scene can be decorrelated using the f/b scheme once on the original array. The GEESE scheme can then be employed on the smoothed array output covariance matrix to estimate the two directions of arrival. Complete the analysis for estimator bias, variance and resolution threshold for the least favorable configuration as well as the most favorable configuration in this case, and compare the results with the uncorrelated case described in the previous problem.

References

[1]   H. L. Van Trees, *Detection, Estimation, and Modulation Theory,* Part I. New York: John Wiley and Sons, 1968.

[2]   V. K. Rohatgi, *An Introduction to Probability Theory and Mathematical Statistics*. New York: John Wiley and Sons, 1976.

[3]   R. J. Bickel and K. A. Doksum, *Mathematical Statistics*. San Francisco, CA: Holden-Day, 1977.

[4]   T. W. Anderson, *An Introduction to Multivariate Statistical Analysis*, 2nd ed. New York: John Wiley and Sons, 1984.

[5]   N. R. Goodman, "Statistical analysis based on a certain multivariate complex Gaussian distribution (an introduction)," *Ann. Math. Stat.*, vol. 34, pp. 152-177, 1963.

[6]   A. B. Baggeroer, "Confidence intervals for regression (MEM) spectral estimates", *IEEE Trans. Inform. Theory*, vol. IT-22, pp. 534-545, Sept. 1976.

[7]   R. P. Gupta, "Asymptotic theory for principal component analysis in the complex case", *J. Indian Stat. Assoc.*, vol. 3, pp. 97-106, 1965.

[8]   R. A. Monzingo and T. W. Miller, *Introduction to Adaptive Arrays*. New York: John Wiley and Sons, 1980.

[9]   I. S. Reed, "On a moment theorem for complex Gaussian processes," *IRE Trans. Inform. Theory*, pp. 194-195, Apr. 1962.

[10]   M. Kaveh and A. J. Barabell, "The statistical performance of the MUSIC and the minimum-norm algorithms in resolving plane waves in noise," *IEEE Trans. Acoust., Speech, Signal Processing*, vol. ASSP-34, pp. 331-341, Apr. 1986.

[11]   S. U. Pillai and B. H. Kwon, "Performance analysis of eigenvector-based high resolution estimators for direction finding in correlated and coherent scenes," in *ONR Annual Report*, Polytechnic University, June 1988.

[12]   M. Okamato, "Distinctness of the eigenvalues of a quadratic form in a multivariate sample," *Annals of Statistics*, vol. 1, pp. 763-765, 1973.

[13]   T. W. Anderson, "Asymptotic theory for principal component analysis," *Ann. Math. Stat.*, vol. 34, pp. 122-148, 1963.

[14]   D. N. Lawley, "Estimation in fact analysis under various initial assumptions," *British Journal of Statistical Psychology*, vol. II, pp. 1-12, 1958.

[15] S. N. Roy, *Some Aspects of Multivariate Analysis*. New York: John Wiley and Sons, 1957.

[16] D. R. Cox and D. V. Hinkley, *Theoretical Statistics*. London, England: Chapman and Hall, 1974.

[17] K. V. Mardia, J. T. Kent, and J. M. Bibby, *Multivariate Analysis*. New York: Academic, 1979.

[18] D. N. Simkins, "Multichannel angle of arrival estimation," Ph.D. Dissertation, Stanford Univ., Stanford, CA, 1980.

[19] R. O. Schmidt, "A signal subspace approach to emitter location and spectral estimation," Ph.D. Dissertation, Stanford Univ., Stanford, CA, Nov. 1981.

[20] G. Bienvenu and L. Kopp, "Optimality of high resolution array processing using the eigensystem approach," *IEEE Trans. Acoust., Speech, Signal Processing*, vol. ASSP-31, pp. 1235-1248, Oct. 1983.

[21] A. M. Bruckstein, T. J. Shan, and T. Kailath, "The resolution of overlapping echos," *IEEE Trans. Acoust., Speech, Signal Processing*, vol. ASSP-33, pp. 1357-1367, Dec. 1985.

[22] H. Akaike, "A new look at the statistical model identification," *IEEE Trans. Automat. Contr.*, vol. AC-19, pp. 716-723, Dec. 1974.

[23] J. Rissanen, "Modeling by shortest data description," *Automatica*, vol. 14, pp. 465-471, 1978.

[24] S. Kullback, J. C. Keegel, and J. H. Kullback, *Topics in Statistical Information Theory*. New York: Springer-Verlag, 1987.

[25] M. Wax and T. Kailath, "Detection of signals by information theoretic criteria," *IEEE Trans. Acoust., Speech, Signal Processing*, vol. ASSP-33, pp. 387-392, Apr. 1985.

[26] L. C. Zhao, P. R. Krishnaiah, and Z. D. Bai, "On detection of the number of signals in presence of white noise," *J. Multivariate Anal.*, vol. 20, pp. 1-25, 1986.

[27] Y. Q. Yin and P. R. Krishnaiah, "On some nonparametric methods for detection of the number of signals," *IEEE Trans. Acoust., Speech, Signal Processing*, vol. ASSP-35, pp. 1533-1538, Nov. 1987.

[28] A. Kshirsagar, *Multivariate Analysis*. New York: Marcel Dekker, 1972.

[29] C. R. Rao, *Linear Statistical Inference and its Applications*, 2nd ed. New York: John Wiley and Sons, 1973.

[30] Y. Lee, "Direction finding from first order statistics and spatial spectrum estimation," Ph.D. dissertation, Polytechnic Univ., Brooklyn, NY, 1988.

[31] I. S. Reed, J. D. Mallet, and L. E. Brennan, "Rapid convergence rate in adaptive arrays," *IEEE Trans. Aerospace Electron. Syst.*, vol. AES-10, pp. 853-863, Nov. 1974.

[32] R. C. Hanumara, "An alternate derivation of the distribution of the conditioned signal-to-noise ratio," *IEEE Trans. Ant. Propag.*, vol. AP-34, pp. 463-464, Mar. 1986.

[33] C. G. Khatri and C. R. Rao, "Effects of estimated noise covariance matrix in optimal signal detection," *IEEE Trans. Acoust., Speech, Signal Processing*, vol. ASSP-35, pp. 671-679, May 1987.

[34] S. U. Pillai and B. H. Kwon, "GEESE (GEneralized Eigenvalues utilizing Signal subspace Eigenvectors) − A new technique for direction finding," *Proc. Twenty Second Annual Asilomar Conference on Signals, Systems, and Computers*, Pacific Grove, CA, Oct. 31 - Nov. 2, 1988.

[35] B. H. Kwon, "New high resolution techniques and their performance analysis for angle-of-arrival estimation," Ph.D. dissertation, Polytechnic Univ., Brooklyn, NY, 1989.

Chapter 4
Estimation of Multiple Signals

4.1 Introduction

Having been able to detect the total number of sources present in the scene, their arrival angles, power levels and crosscorrelations, often one may be specifically interested in the actual signal of one of these targets. Letting $d(t)$ denote this desired signal $u_{k_0}(t)$, the next objective is to attempt to estimate the actual waveform $d(t)$ associated with $u_{k_0}(t)$ by improving the overall reception of $d(t)$ in an environment having several sources. To achieve this, ideally one must be able to suppress the undesired signals and enhance the desired signal. The desired signal may correspond to a friendly satellite signal in presence of hostile jammers, and in this case the second order characteristics themselves may change with time. This can happen because of physical motion or deliberate on-off jamming strategies of the smart opponent. In this case, capabilities for quick adaptive learning of the changing scene are required to maintain an acceptable level of the desired signal characteristics at the receiver.

At times, the desired signal structure might be partially known and the objective in that case is to detect its presence in the available noisy data. This situation is often encountered in sonar where the data is analyzed at the receiver to detect the presence of the signature of a specific class of submarine. Though the signal structure is known, it may still contain unknown parameters such as angle of arrival or random phase.

In all these cases, a reasonable strategy is to find the best set of weights to be used at the sensor outputs for combining them optimally under some suitable criterion. Minimization of mean square error (MMSE), maximization of SNR (MSNR), ML criterion have been widely used as suitable optimization criteria for radar, sonar and communication systems, and in all these cases the optimal weights turn out to be a function of signal strengths – desired and undesired –, their directions of arrival, and noise variance/covariance characteristics.

Such optimal solutions, even if hard to realize physically, can be useful in establishing valuable upper bounds on their steady state performance. These bounds in turn can be used to compare the relative efficiency of closely related physically realizable solutions. We will examine these various optimality measures to determine the respective steady state solutions for the weight vector in narrowband situations and analyze their physically realizable counterparts.

4.2 Optimum Processing : Steady State Performance and the Wiener Solution

As before let $\mathbf{x}(t)$ represent the output vector at an M element array that receives signals from the desired as well as undesired narrowband sources in presence of noise. Further, let $d(t)$ denote the desired signal and $\mathbf{w}$ as given by (2.18) represent the desired weight vector for linearly combining the array outputs. From Fig. 2.3 the combined array output $\hat{d}(t)$ that approximates $d(t)$ is given by (2.17), i.e.,

$$\hat{d}(t) = \mathbf{w}^{\dagger}\mathbf{x}(t). \tag{4.1}$$

A similar solution to a wideband situation is not difficult to realize. In fact, if $\mathbf{w}(\theta)$ represents the optimum weight vector in a narrowband situation, then comparing (2.10) and (2.12), under the same criterion, the wideband case gives rise to a frequency dependent weight vector $\mathbf{w}(f, \theta)$, i.e., a set of linear filters that usually can be approximated by a tapped delay line at each sensor output [1, 2]. We begin with the mean square error (MSE) performance measure for a narrowband scene.

Minimization of Mean Square Error (MMSE)

The MSE performance measure was considered by Widrow *et al.* [3] and subsequently extensions based on this criterion have been developed [2, 4]. As the name implies, the MSE measure minimizes the error $e(t)$ defined by the difference between the desired signal and the actual array response $\hat{d}(t)$. Thus

$$e(t) = d(t) - \mathbf{w}^{\dagger}\mathbf{x}(t) \tag{4.2}$$

and the mean square error ε is given by

$$\varepsilon(\mathbf{w}) = E[\,|e(t)|^2\,]$$

$$= E[\,|d(t)|^2\,] - 2\,\mathrm{Re}(\mathbf{w}^\dagger E[\mathbf{x}(t)d^*(t)]) + \mathbf{w}^\dagger \mathbf{R}\mathbf{w}$$

$$= P - \mathrm{Re}(2\mathbf{w}^\dagger \boldsymbol{\gamma}_{xd} - \mathbf{w}^\dagger \mathbf{R}\mathbf{w}) \tag{4.3}$$

where $P = E[\,|d(t)|^2\,]$ represents the desired signal power and

$$\boldsymbol{\gamma}_{xd} = E[\mathbf{x}(t)d^*(t)] \tag{4.4}$$

represents the crosscorrelation between the array output vector and the desired signal. The minimum of the quadratic function in (4.3) can be found by setting its gradient $\nabla_{\mathbf{w}}(\varepsilon)$ with respect to $\mathbf{w}$ equal to zero. Since

$$\nabla_{\mathbf{w}}(\varepsilon) = -2\boldsymbol{\gamma}_{xd} + 2\mathbf{R}\mathbf{w}, \tag{4.5}$$

the optimum choice for the weight vector must satisfy

$$\nabla_{\mathbf{w}}(\varepsilon) = -2(\mathbf{R}\mathbf{w} - \boldsymbol{\gamma}_{xd}) = 0$$

or

$$\mathbf{w}_{opt} = \mathbf{R}^{-1}\boldsymbol{\gamma}_{xd}. \tag{4.6}$$

This is the Wiener-Hopf equation in matrix form and is often referred to as the optimum Wiener solution [1]. With (4.6) in (4.3), the resulting minimum MSE is found to be

$$\varepsilon_{\min} = P - \boldsymbol{\gamma}_{xd}^\dagger \mathbf{R}^{-1}\boldsymbol{\gamma}_{xd}. \tag{4.7}$$

Maximization of Signal-to-Noise Ratio (MSNR)

The array weight vector can also be optimized by maximizing the output SNR with respect to the desired signal. This is carried out by considering $\mathbf{x}(t)$ to consist of the desired component $\mathbf{u}(t) = d(t)\mathbf{a}(\omega)$, where $\mathbf{a}(\omega)$ represents the direction vector associated with the desired source, and the undesired component $\mathbf{n}(t)$ so that

$$\mathbf{x}(t) = d(t)\mathbf{a}(\omega) + \mathbf{n}(t) \triangleq \mathbf{u}(t) + \mathbf{n}(t). \tag{4.8}$$

This gives the output SNR to be

$$(SNR)_o = \frac{E[\ |\mathbf{w}^\dagger d(t)\mathbf{a}(\omega)|^2\ ]}{E[\ |\mathbf{w}^\dagger \mathbf{n}(t)|^2\ ]} = \frac{P\ |\mathbf{w}^\dagger \mathbf{a}(\omega)|^2}{\mathbf{w}^\dagger \mathbf{R}_n \mathbf{w}} \qquad (4.9)$$

where

$$\mathbf{R}_n = E[\mathbf{n}(t)\mathbf{n}^\dagger(t)] \qquad (4.10)$$

is the noise covariance matrix. For nonsingular $\mathbf{R}_n$, using Schwarz' inequality in (4.9) we have

$$(SNR)_o = \frac{P\ |(\mathbf{R}_n^{1/2}\mathbf{w})^\dagger(\mathbf{R}_n^{-1/2}\mathbf{a}(\omega))|^2}{\mathbf{w}^\dagger \mathbf{R}_n \mathbf{w}}$$

$$\leq P\,\mathbf{a}^\dagger(\omega)\,\mathbf{R}_n^{-1}\,\mathbf{a}(\omega) \triangleq (SNR)_{\max}. \qquad (4.11)$$

An easy verification shows that equality is achieved in (4.11) with

$$\mathbf{w}_{SNR} = k\,\mathbf{R}_n^{-1}\,\mathbf{a}(\omega) \qquad (4.12)$$

where k is any nonzero complex constant. Thus the optimum weight vector that maximizes the output SNR of the desired signal is given by (4.12), which is also seen to be the Wiener solution given by (4.6) provided the desired signal is uncorrelated with the rest of data (see problem 2).

Designing the processor so that the weights satisfy (4.12) (i.e., $\mathbf{R}_n \mathbf{w} = k\,\mathbf{a}(\omega)$) implies that output SNR is the governing performance criterion, even when no jamming signal and no desired signal is present. This is because inherent in itself, this criterion has no mechanism to tell whether the random process $\hat{d}(t)$ at the processor output consists of the desired signal $d(t)$ obscured by noise or consists of noise alone. To realize such goals, one must concentrate on mechanisms that minimize the risk associated with incorrect decisions or maximize the probability of detection. The likelihood ratio test (LRT) is optimum in such hypothesis testing situations for a wide range of problems [5].

In a binary hypothesis testing situation, with H_1 representing "signal is present in the data" and H_0 "signal is absent in the data", the LRT is given by

$$L(\mathbf{x}) \triangleq \frac{\text{conditional p.d.f. of the observations given signal is present}}{\text{conditional p.d.f. of observations in noise}}$$

$$\triangleq \frac{f_{\mathbf{x}|H_1}(\mathbf{x}|H_1)}{f_{\mathbf{x}|H_0}(\mathbf{x}|H_0)} \tag{4.13}$$

and decisions are made in favor of H_1 or H_0 by comparing $L(\mathbf{x})$ with a known threshold η as follows

$$L(\mathbf{x}) \underset{H_0}{\overset{H_1}{\gtrless}} \eta. \tag{4.14}$$

If the likelihood ratio test depends on an unknown parameter θ, whose p.d.f. is known, then the best test is the optimum Bayes test [6]. This test is given by averaging $L(\mathbf{x}|\theta)$ with respect to θ and comparing the result with the given threshold η. That is

$$L(\mathbf{x}) = \int L(\mathbf{x}|\theta) f_\theta(\theta) d\theta \underset{H_0}{\overset{H_1}{\gtrless}} \eta \tag{4.15}$$

where

$$L(\mathbf{x}|\theta) = \frac{f_{\mathbf{x}|H_1}(\mathbf{x}|\theta, H_1)}{f_{\mathbf{x}|H_0}(\mathbf{x}|\theta, H_0)}. \tag{4.16}$$

Often $L(\mathbf{x})$ may turn out to be a monotonic function of a scalar variable $l(\mathbf{x})$, known as the sufficient statistic, which contains all necessary information required to make a decision in favor of either H_1 or H_0. In terms of the sufficient statistic, (4.14) can be rewritten as

$$l(\mathbf{x}) \underset{H_0}{\overset{H_1}{\gtrless}} \gamma$$

where γ is the new threshold and this gives the probability of detection P_D and the probability of false alarm P_F to be [6]

$$P_D \triangleq P(H_1|H_1) = P(l \geq \gamma|H_1) = \int_\gamma^\infty f_{l|H_1}(l|H_1) dl \tag{4.17}$$

and

$$P_F \triangleq P(H_1|H_0) = P(l \geq \gamma|H_0) = \int_{\gamma}^{\infty} f_{l|H_0}(l|H_0)\,dl \ . \quad (4.18)$$

In radar and sonar, P_F is preset to be a small constant that the system can tolerate, and from (4.18) this determines γ, which in turn determines P_D.

Brennan and Reed have shown that for a nonrandom desired signal in presence of Gaussian noise, the above formulation leads to maximization of output SNR as the optimal choice for minimizing the risk associated with incorrect decisions [7]. This in turn provides a firmer statistical ground for the maximization of the SNR procedure. To see this, from (4.8) we have

$$\mathbf{x} = \begin{cases} \mathbf{u} + \mathbf{n} & H_1 \\ \mathbf{n} & H_0 \end{cases}$$

where $\mathbf{u}$ is unknown, nonrandom, and $\mathbf{n}$ is M-variate zero mean complex Gaussian signal. The processor output $z = \mathbf{w}^\dagger \mathbf{x}$ is also complex Gaussian with mean

$$E[z] = \begin{cases} \mathbf{w}^\dagger \mathbf{u} & H_1 \\ 0 & H_0 \end{cases}$$

and common variance

$$\sigma^2 = E[\,|z - \mathbf{w}^\dagger \mathbf{u}|^2\,] = \mathbf{w}^\dagger \mathbf{R}_n \mathbf{w} \ .$$

This gives

$$f_{z|H_1}(z|H_1) = \frac{1}{\pi \sigma^2} e^{-|z - \mathbf{w}^\dagger \mathbf{u}|^2/\sigma^2}$$

$$f_{z|H_0}(z|H_0) = \frac{1}{\pi \sigma^2} e^{-|z|^2/\sigma^2}$$

and with this in (4.13) and (4.16) the optimum LRT decision criterion reduces to

$$L = e^{\frac{1}{\sigma^2}\left(2a\,|z|\,\cos(\theta-\delta) - a^2\right)}$$

$$\quad (4.19)$$

where θ is the phase angle associated with z, and further, a and δ represent the magnitude and phase associated with $\mathbf{w}^{\dagger}\mathbf{u}$, i.e.,

$$z = |z| e^{j\theta}; \quad \mathbf{w}^{\dagger}\mathbf{u} = a e^{j\delta}, \quad a = |\mathbf{w}^{\dagger}\mathbf{u}|. \tag{4.20}$$

Notice that the phase angle θ of the received signal is a random variable and, in the absence of any other information, it may be reasonably approximated as uniformly distributed in the interval $(0, 2\pi)$. With (4.19) in (4.15), this gives the optimum Bayes test to be

$$L(z) = e^{-a^2/\sigma^2} \int_0^{2\pi} e^{2a\,|z|\,\cos(\theta-\delta)/\sigma^2}\, \frac{d\theta}{2\pi}$$

$$= e^{-a^2/\sigma^2} I_0\left(\frac{a\,|z|}{\sigma^2}\right) \underset{H_0}{\overset{H_1}{\gtrless}} \eta \tag{4.21}$$

where $I_0(\cdot)$ is the modified zero order Bessel function of the first kind. Since $I_0(\cdot)$ is an increasing function of its argument and η is a positive constant, the above test is equivalent to

$$l(z) = |z| \underset{H_0}{\overset{H_1}{\gtrless}} \gamma$$

where $l(z)$ represents the sufficient statistic and γ satisfies

$$e^{-a^2/\sigma^2} I_0\left(\frac{a\gamma}{\sigma^2}\right) = \eta.$$

The probability of false alarm P_F and the probability of detection P_D can be obtained easily from these relations. In fact since $|z|^2$ is exponential with parameter σ^2 under the null hypothesis H_0 and has a noncentral $\chi^2(2)$ distribution under H_1, we have

$$P_F = P(l > \gamma | H_0) = P(|z|^2 > \gamma^2 | \mathbf{x} = \mathbf{n}) = e^{-\gamma^2/\sigma^2}$$

and

$$P_D = P(l > \gamma \,|\, H_1) = P(\,|z|^2 > \gamma^2 \,|\, H_1)$$

$$= \frac{1}{\sigma^2} \int_\gamma^\infty I_0\left(\frac{2\gamma a}{\sigma^2}\right) e^{(-\gamma^2 + a^2)/2\sigma^2} \gamma \, d\gamma. \qquad (4.22)$$

For a fixed P_F, with $x = \gamma/\sigma$ and

$$\alpha^2 = \frac{a^2}{\sigma^2} = \frac{|\mathbf{w}^\dagger \mathbf{u}|^2}{\mathbf{w}^\dagger \mathbf{R}_n \mathbf{w}}$$

representing the output SNR, (4.22) reduces to

$$P_D = \int_{\frac{\gamma}{\sigma} = \sqrt{\ln(1/P_F)}}^{\infty} I_0(2\alpha x) e^{(-x^2 + \alpha^2)} x \, dx \,.$$

Thus for a given P_F, the probability of detection is a function of the output SNR α^2. Moreover, with $\beta = \sqrt{\ln(1/P_F)}$ the derivative of $P_D(\alpha)$ can be shown to be [7]

$$\frac{dP_D(\alpha)}{d\alpha} = \beta e^{-(\beta^2 + \alpha^2)} I_1(\alpha\beta) > 0$$

since $I_1(x) > 0$ for all $x > 0$. Thus $P_D(\alpha)$ is a monotone increasing function of α, implying that to maximize the detection probability with respect to the weight vector $\mathbf{w}$, it is sufficient to maximize the output SNR α^2. This of course can be realized by the weight vector $\mathbf{w}_{SNR}$ given in (4.12).

Maximum Likelihood Performance Measure

In this case the desired signal is unknown, and for the received signal given by (4.8), under the assumption that the noise components have multivariate Gaussian distribution, from (3.3) the likelihood function is given by

$$L = f(\mathbf{x}(t)) = \frac{1}{|\pi \mathbf{R}_n|} e^{-[\mathbf{x}(t) - d(t)\mathbf{a}(\omega)]^\dagger \mathbf{R}_n^{-1}[\mathbf{x}(t) - d(t)\mathbf{a}(\omega)]} \,. \qquad (4.23)$$

The maximum likelihood processor is obtained by solving for the estimate $\hat{d}(t)$ of $d(t)$, that maximizes (4.23) or equivalently its logarithm. Equating the partial derivative of $ln\ L$ with respect to $d(t)$, this yields

$$\hat{d}(t)\,\mathbf{a}^{\dagger}(\omega)\,\mathbf{R}_n^{-1}\,\mathbf{a}(\omega) = \mathbf{a}^{\dagger}(\omega)\,\mathbf{R}_n^{-1}\,\mathbf{x}(t)$$

or

$$\hat{d}(t) = \frac{(\mathbf{R}_n^{-1}\,\mathbf{a}(\omega))^{\dagger}\,\mathbf{x}(t)}{\mathbf{a}^{\dagger}(\omega)\,\mathbf{R}_n^{-1}\,\mathbf{a}(\omega)} = \mathbf{w}_{ML}^{\dagger}\,\mathbf{x}(t). \qquad (4.24)$$

Thus, the optimum weight vector in this case has the form

$$\mathbf{w}_{ML} = \frac{(\mathbf{R}_n^{-1}\,\mathbf{a}(\omega))}{\mathbf{a}^{\dagger}(\omega)\,\mathbf{R}_n^{-1}\,\mathbf{a}(\omega)} = k'\,\mathbf{R}_n^{-1}\,\mathbf{a}(\omega) \qquad (4.25)$$

and a quick comparison with (4.12) shows that the ML solution also maximizes the SNR. Notice that when the desired signal is uncorrelated with the remaining set of signals, the MMSE solution also is a scalar multiple of (4.12) (see problem 2). This is only natural, since for Gaussian data the MMSE criterion and the ML criterion lead to the same estimator [6].

Broadband Processing

In this case, the received signal in a single broadband source scene can be expressed as

$$\mathbf{x}(t) = \int_{-\infty}^{t} \mathbf{a}(t-\tau, \theta)\,u(\tau)\,d\tau + \mathbf{n}(t) \qquad (4.26)$$

where $\mathbf{a}(t,\theta)$ is a known linear transformation that represents propagation effects and signal distortions that occur in the sensors. When the broadband signal $u(t)$ and noise $\mathbf{n}(t)$ are stationary processes and the observation interval is long, then the convolution appearing in (4.26) can be avoided by Fourier transform techniques. Thus transforming (4.26) into the frequency domain by taking its Fourier transform, we obtain formally

$$x(f) = a(f, \theta)u(f) + n(f). \tag{4.27}$$

Although (4.27) is in the frequency domain, it has the same form as (4.P.1) in the time domain. This implies that the MMSE estimator for the unknown random signal $u(f)$ in (4.27) is structurally similar to (4.P.2) with covariances replaced by cross spectral density matrices. Thus the frequency domain equivalent of (4.P.2) with $\mu = 0$ gives the desired solution

$$\hat{u}(f) = \Phi_n(f)a^\dagger(f)\left[a^\dagger(f)\Phi_u(f)a(f) + \Phi_n(f)\right]x(f). \tag{4.28}$$

Here the cross spectral densities $\Phi_u(f)$ and $\Phi_n(f)$ are the Fourier transforms of covariance matrices $R_u(\tau)$ and $R_n(\tau)$, respectively. Using the matrix identity

$$\mathbf{PM}^\dagger(\mathbf{MPM}^\dagger + \mathbf{Q})^{-1} = (\mathbf{P}^{-1} + \mathbf{M}^\dagger\mathbf{Q}^{-1}\mathbf{M})^{-1}\mathbf{M}^\dagger\mathbf{Q}^{-1}, \tag{4.29}$$

we get the more useful form

$$\hat{u}(f) = \left[\Phi_u^{-1}(f) + a^\dagger(f)\Phi_n^{-1}(f)a(f)\right]^{-1} a^\dagger(f)\Phi_n^{-1}(f)x(f). \tag{4.30}$$

Since the quantity $\Phi_u^{-1}(f) + a^\dagger(f)\Phi_n^{-1}(f)a(f)$ is a positive scalar, (4.30) can be rewritten as

$$\hat{u}(f) = |m(f)|^2 h(f)x(f) \tag{4.31}$$

where

$$|m(f)|^2 = \frac{\Phi_u(f)}{1 + \Phi_u(f)a^\dagger(f)\Phi_n^{-1}(f)a(f)}$$

and

$$h(f) = a^\dagger(f)\Phi_n^{-1}(f). \tag{4.32}$$

In general, the frequency response in (4.31) is unrealizable and it is necessary to introduce a time delay to obtain a good approximation to the corresponding time waveform given by

$$\hat{u}(t) = \frac{1}{2\pi} \int\limits_{-\infty}^{\infty} |m(f)|^2 h(f)x(f)e^{j2\pi ft}\, df . \tag{4.33}$$

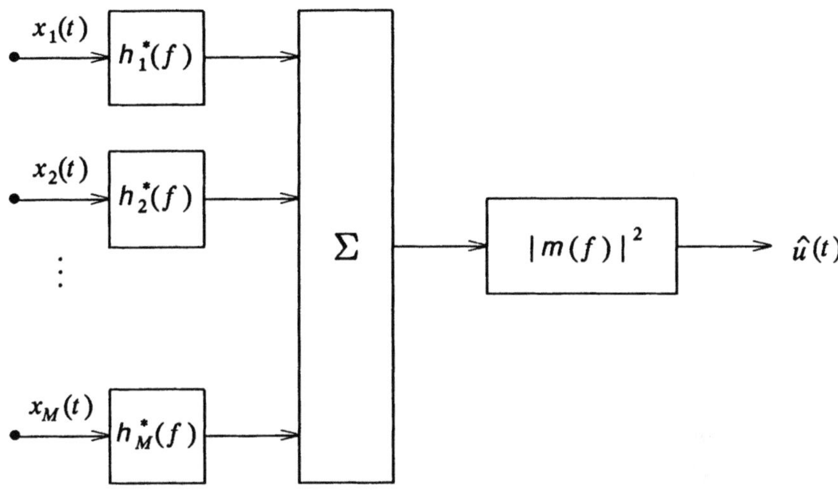

Fig. 4.1 Optimum array processor for estimation of a broadband signal

The filter $h(f)$ can be given an interesting physical interpretation: From (4.32) it first performs the operation of spatial prewhitening and then matches to the known propagation and distortion effects represented by $a(f)$. Thus, as illustrated in Fig. 4.1, each filter $h_i(f)$ employs a prewhitening followed by a matched filtering operation on the received signal $x_i(t)$, $i = 1, 2, \cdots, M$.

4.3 Implementation of the Wiener Solution

A variety of popular performance measures have been shown to result in solutions that are closely related to the Wiener solution. As seen from (4.6), to implement the Wiener solution directly, the complete knowledge of $\mathbf{R}$ and γ_{xd} is required and usually these quantities are unknown. In addition, the inversion of $\mathbf{R}$ is an expensive and time consuming operation, especially for large arrays. Further, as remarked earlier, the source scene may be changing with time, and in that case, the time varying output covariance matrix $\mathbf{R}(t)$ needs to be known for implementing the Wiener solution directly. This motivates the need for a recursive solution for the weight vector that is simple to implement and is adaptive to the changing environment [8].

At any stage, the weight adjustment factor should depend on the performance measure under consideration and the present weight

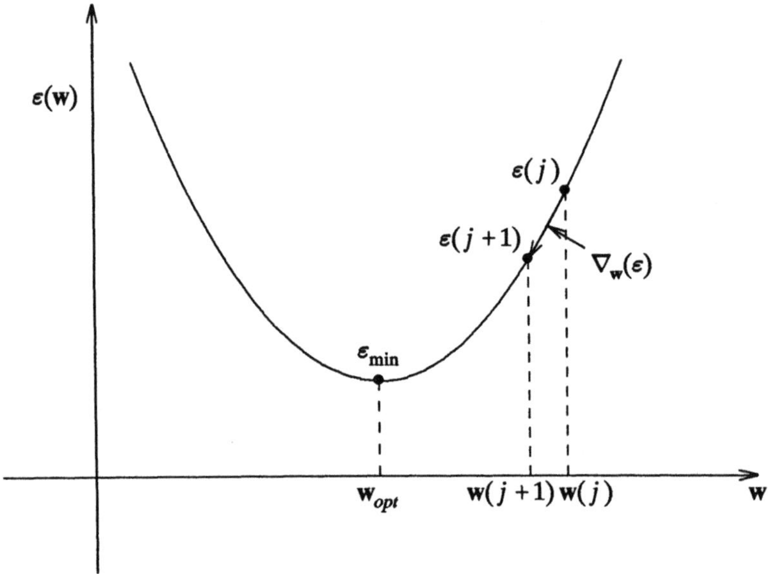

Fig. 4.2 Steepest descent in a quadratic bowl.

vector. For any performance measure, there exist(s) (an) optimum weight vector(s) and the recursion should preferably take place "along the shortest path available" towards this desired solution. Under the quadratic performance measure of minimizing the mean square error, this performance measure can be visualized as a bowl-shaped surface so that the adaptive processor has the task of continually seeking the "bottom of the bowl." In this case, the shortest path can be realized by "hill climbing" methods of which the various gradient based algorithms such as steepest descent, least mean square (LMS) and accelerated gradient methods are representative. The gradient based techniques can be best explained in terms of the steepest descent method.

4.3.1 The Method of Steepest Descent

Under the MMSE criterion, the optimum solution for the weight vector is given by $\mathbf{w}_{opt}$ in (4.6) and the associated residual error is given by $\varepsilon_{\min}$ in (4.7). This $\varepsilon_{\min}$ represents the bottom of the quadratic bowl generated by $\varepsilon(\mathbf{w})$ in (4.3) (see Fig. 4.2). Any other weight vector $\mathbf{w}$ generates $\varepsilon(\mathbf{w}) \geq \varepsilon_{\min}$, and hence at the jth stage of recursion the present weight vector $\mathbf{w}(j)$ should change to $\mathbf{w}(j+1)$ such that the respective mean square errors satisfy $\varepsilon(\mathbf{w}(j+1)) \leq \varepsilon(\mathbf{w}(j))$. The fastest mode to accomplish this change is along the direction of the steepest descent, which is opposite to the direction of the gradient of $\varepsilon(\mathbf{w})$. This is only natural, since the gradient $\nabla_{\mathbf{w}}(\varepsilon)$ of $\varepsilon(\mathbf{w})$ with respect to $\mathbf{w}$ represents the direction along which $\varepsilon(\mathbf{w})$ has the maximum increment. With Δ_s denoting the step size of descent, the above arguments can be summarized into the compact form

$$\mathbf{w}(j+1) = \mathbf{w}(j) - \Delta_s \, \nabla_{\mathbf{w}}(\varepsilon) \, . \tag{4.34}$$

Substituting (4.5) in (4.34) yields

$$\mathbf{w}(j+1) = \mathbf{w}(j) - 2\Delta_s \, (\mathbf{R}\mathbf{w}(j) - \boldsymbol{\gamma}_{xd}) \, . \tag{4.35}$$

The method of steepest descent begins with an initial guess of the weight vector $\mathbf{w}(0)$ and this together with the current gradient vector determines the weight vector at the next stage.

A quick glance at (4.35) reveals that this recursive algorithm is still impractical since it makes use of the signal environment statistics $\mathbf{R}$ and $\boldsymbol{\gamma}_{xd}$ that are usually unknown. In a stationary environment, the LMS algorithm circumvents this problem by making use of the ML-estimate of the unknown gradient vector in its implementation. Consequently performance limits on the steepest descent algorithm also act as an upper bound on the LMS algorithm. To analyze these restrictions it is instructive to uncouple the M equations in (4.35) by diagonalizing the covariance matrix $\mathbf{R}$. From (2.60) we have

$$\mathbf{R} = \mathbf{B}\boldsymbol{\Lambda}\mathbf{B}^{\dagger} \, ; \quad \mathbf{B} = \left[\boldsymbol{\beta}_1, \boldsymbol{\beta}_2, \cdots, \boldsymbol{\beta}_M\right] \, , \quad \mathbf{B}\mathbf{B}^{\dagger} = \mathbf{I} \tag{4.36}$$

with $\boldsymbol{\Lambda} = diag\,[\lambda_1, \lambda_2, \cdots, \lambda_M]$. Premultiplying (4.35) with $\mathbf{B}^{\dagger}$, it can be rewritten as

$$\tilde{\mathbf{w}}(j+1) = \tilde{\mathbf{w}}(j) - 2\Delta_s \, (\boldsymbol{\Lambda}\tilde{\mathbf{w}}(j) - \tilde{\boldsymbol{\gamma}}_{xd}) \tag{4.37}$$

where by definition

$$\tilde{\mathbf{w}}(j+k) = \mathbf{B}^{\dagger}\tilde{\mathbf{w}}(j+k) \, , \quad k = 0, 1$$

and

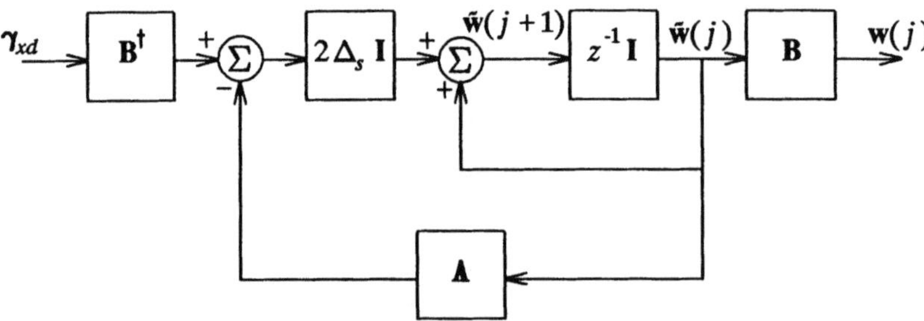

Fig. 4.3 Block representation of steepest descent method.

$$\tilde{\gamma}_{xd} = \mathbf{B}^{\dagger}\gamma_{xd} \; .$$

A block representation of (4.37) gives rise to the feedback model in Fig. 4.3 in which all previously existed cross-couplings have been eliminated. The transients of each of the M paths in Fig. 4.3 can be analyzed by considering the behavior of a typical loop. The transfer function of the isolated closed loop feedback system corresponding to the kth coordinate has the form

$$\frac{\tilde{w}_k(z)}{\gamma_{xd,k}(z)} = \frac{\Delta_s z^{-1}}{1 - (1 - 2\Delta_s \lambda_k)z^{-1}} \tag{4.38}$$

where λ_k is the kth diagonal entry in $\mathbf{A}$. The impulse response of (4.38) can be obtained by letting $\gamma_{xd,k}(z) = 1$ and taking the inverse z-transform of the resulting output. This gives

$$\tilde{w}_k(n) = \Delta_s (1 - 2\Delta_s \lambda_k)^{n-1}, \quad n \geq 0 .$$

Clearly, the above time response represents a stable system so long as

$$| 1 - 2\Delta_s \lambda_k | < 1 .$$

Since $\lambda_k > 0$, the above restriction is equivalent to

$$\Delta_s > 0 \quad \text{and} \quad |\lambda_k \Delta_s | < 1$$

or

$$0 < \Delta_s < \frac{1}{\lambda_k}, \quad \text{all } k.$$

Consequently the stability of the multidimensional flow graph in Fig. 4.3 and hence that of the steepest descent adaptation process is guaranteed if and only if

$$0 < \Delta_s < \frac{1}{\lambda_{\max}} \tag{4.39}$$

where $\lambda_{\max}$ is the largest eigenvalue of $\mathbf{R}$. In an uncorrelated source scene since

$$\lambda_{\max} \leq \sum_{k=1}^{M} \lambda_k \leq M \sum_{k=1}^{K} P_k + \sigma^2 \triangleq P_{IN}$$

the above stability requirement is satisfied so long as

$$0 < \Delta_s < \frac{1}{P_{IN}}.$$

4.3.2 The Least Mean Square (LMS) Algorithm

In a stationary environment, when the statistics describing the source scene are unknown, exact computation of the gradient of the performance surface is no longer possible. Under these conditions, the LMS algorithm introduced by Widrow replaces the unknown gradient in (4.34) by its estimate and has proved to be particularly useful. Consequently the LMS algorithm is exactly like the steepest descent method except that changes in the weight vector are made in the direction given by an estimated gradient vector rather than the actual gradient vector. Thus the corresponding recursive equation for the weight vector can be expressed as

$$\mathbf{w}(j+1) = \mathbf{w}(j) - \Delta_s \, \hat{\nabla}_\mathbf{w}[\varepsilon(j)] \tag{4.40}$$

where the MSE, $\varepsilon(j) \triangleq \varepsilon[\mathbf{w}(j)]$ is as given in (4.3). The required estimated gradient of the MSE, $\hat{\nabla}_\mathbf{w}[\varepsilon(j)]$ can be obtained from the gradient of the estimate of $\varepsilon(j)$ computed from a single time sample of the squared error, i.e.,

$$\hat{\nabla}_{\mathbf{w}}[\varepsilon(j)] = \nabla_{\mathbf{w}}[\hat{\varepsilon}(j)]$$

where

$$\hat{\varepsilon}(j) = |e(j)|^2 \qquad (4.41)$$

and from (4.2)

$$e(j) = d(j) - \mathbf{w}^{\dagger}(j)\mathbf{x}(j). \qquad (4.42)$$

This gives

$$\nabla_{\mathbf{w}}[\hat{\varepsilon}(j)] = 2e^{*}(j)\nabla_{\mathbf{w}}[e(j)] = -2e^{*}(j)\mathbf{x}(j)$$

so that the estimated gradient becomes

$$\hat{\nabla}_{\mathbf{w}}[\varepsilon(j)] = -2e^{*}(j)\mathbf{x}(j). \qquad (4.43)$$

Notice that

$$E[\hat{\nabla}_{\mathbf{w}}[\varepsilon(j)]] = 2(\mathbf{R}\mathbf{w}(j) - \boldsymbol{\gamma}_{xd}) = \nabla_{\mathbf{w}}[\varepsilon(j)]$$

implying that the gradient estimator is unbiased. Substituting (4.43) in (4.40) we get the new iterative relation

$$\mathbf{w}(j+1) = \mathbf{w}(j) + 2\Delta_{s}\, e^{*}(j)\mathbf{x}(j) \qquad (4.44)$$

which represents the discrete form of the LMS algorithm. Thus the old weight vector is updated by the current input signal vector that has been scaled by the conjugate of the error signal.

From (4.44), the LMS algorithm requires the error signal given by (4.42) for its implementation and such an error signal in turn requires a reference signal model $d(j)$ to represent the signal it is desired to receive. As a practical matter, the desired signal is usually not available for LMS adaptation process. This is indeed the case in radar and sonar systems where the desired signal is either absent or is an unknown random process. In fact if it were available by some other means, there would be no need for the receiving array and the processor. This serious drawback can be traced back to the steepest descent algorithm, where knowledge of $\mathbf{R}$ and $\boldsymbol{\gamma}_{xd}$ are required for its implementation. However, in communication systems, where the desired signal is usually present or at least is known in terms of its structure, the above requirement can be met by generating an approximation to the actual signal. Many practical communication systems employing the LMS algorithm derive this reference signal from the array output, and this requires some knowledge of the directional and spectral characteristics of the incoming signal of interest. At times this information may be known a priori, or can be estimated. If the distinguishing features of the desired and interference signals are available, then these known differences can be used to isolate these two signal classes. In particular, if the desired signal is known to be uncorrelated with the interfering signals, the reference signal need not be a perfect replica of the desired one; instead it is sufficient that the reference signal be highly correlated with the desired signal. This will guarantee that the adaptation process will behave in the desired manner, since it is only the correlation between the reference signal and the actual sensor outputs that affects the adaptive weights [2].

Convergence of the LMS Algorithm

To justify the notion that the LMS algorithm is a practical means of implementing the Wiener solution, it is necessary to demonstrate its convergence to the optimum solution given by (4.6). For asymptotic mean square convergence, it is sufficient that, as $j \rightarrow \infty$

$$E[\mathbf{w}(j)] \rightarrow \mathbf{w}_{opt} \tag{4.45}$$

and

$$Cov[\mathbf{w}(j)] \rightarrow \mathbf{O}. \tag{4.46}$$

It will now be shown that in presence of independent observations, the weight vector only satisfies (4.45) and not (4.46). Toward this purpose assume that the time between successive iterations of the LMS algorithm is long enough so that the data $\mathbf{x}(0)$, $\mathbf{x}(1)$, $\cdots$ are independent of each other. From (4.44) it follows that $\mathbf{w}(j)$ is a function only of $\mathbf{x}(j\text{-}1)$, $\mathbf{x}(j\text{-}2)$, $\cdots$, $\mathbf{x}(0)$ and $\mathbf{w}(0)$, so that $\mathbf{w}(j)$ is independent of $\mathbf{x}(j)$. Taking expected value of both sides of (4.44) results in

$$E[\mathbf{w}(j+1)] = E[\mathbf{w}(j)] + 2\Delta_s E[\mathbf{x}(j)(d^*(j) - \mathbf{x}^t(j)\mathbf{w}(j))]$$

$$= E[\mathbf{w}(j)] + 2\Delta_s (\gamma_{xd} - \mathbf{R}E[\mathbf{w}(j)])$$

$$= [\mathbf{I} - 2\Delta_s \mathbf{R}]E[\mathbf{w}(j)] + 2\Delta_s \gamma_{xd}.$$

Completing the above iteration up to the starting point $\mathbf{w}(0)$, we have

$$E\left[\mathbf{w}(j+1)\right] = \left[\mathbf{I} - 2\Delta_s \mathbf{R}\right]^{j+1}\mathbf{w}(0) + 2\Delta_s \sum_{i=0}^{j} \left[\mathbf{I} - 2\Delta_s \mathbf{R}\right]^{i} \boldsymbol{\gamma}_{xd} \quad . (4.47)$$

Diagonalization of $\mathbf{R}$ as in (4.36), further simplifies (4.47) into

$$E\left[\mathbf{w}(j+1)\right] = \mathbf{B}\left[\mathbf{I} - 2\Delta_s \mathbf{\Lambda}\right]^{j+1}\mathbf{B}^{\dagger}\mathbf{w}(0)$$

$$+ 2\Delta_s \mathbf{B} \sum_{i=0}^{j} \left[\mathbf{I} - 2\Delta_s \mathbf{\Lambda}\right]^{i} \mathbf{B}^{\dagger}\boldsymbol{\gamma}_{xd} \quad . \qquad (4.48)$$

If the step size in this case also satisfies the stability requirement (4.39) obtained for the steepest descent method, then all the terms in the diagonal matrix $[\mathbf{I} - 2\Delta_s \mathbf{\Lambda}]$ have magnitude less than unity, and hence

$$\lim_{j \to \infty} \left[\mathbf{I} - 2\Delta_s \mathbf{\Lambda}\right]^{j+1} \to \mathbf{O}. \qquad (4.49)$$

As a result, the first term in (4.48) vanishes after a large number of iterations. Moreover, the summation factor in the second term becomes

$$\lim_{j \to \infty} \sum_{i=0}^{j} \left[\mathbf{I} - 2\Delta_s \mathbf{\Lambda}\right]^{i} = \frac{1}{2\Delta_s} \mathbf{\Lambda}^{-1}$$

and consequently, after a sufficient number of iterations, (4.48) gives

$$\lim_{j \to \infty} E\left[\mathbf{w}(j+1)\right] = 2\Delta_s \mathbf{B}\left(\frac{1}{2\Delta_s} \mathbf{\Lambda}^{-1}\right)\mathbf{B}^{\dagger}\boldsymbol{\gamma}_{xd}$$

$$= \mathbf{R}^{-1}\boldsymbol{\gamma}_{xd} = \mathbf{w}_{opt}. \qquad (4.50)$$

This result is the same as (4.45). The foregoing convergence result assumes that successive input data samples are independent and that the step size obeys the same restrictions set forth in the steepest descent method. Griffiths has presented experimental evidence indicating that the above independence assumption is overly restrictive [9].

From (4.48) – (4.50), the speed of adaptation to the Wiener solution will depend upon the step size and the number of data samples used to compute the statistical averages. If a large step size is used, then the excursions in adjacent weight values will also tend to be large, with the result that many iterations are required before weight values become acceptably close to the desired solution. A small step size, on the other hand, though results in smooth transitions of the weight vector, may require unacceptably large number of iterations for convergence. Similarly if a small number of samples is used to estimate the desired statistical averages, then the adaptation will be fast, but the quality of the resulting weight vectors and the expected steady state performance will not be high. In that case the actual MSE will exceed the minimum MSE ε_{min} and their difference can be used as a measure of the extent to which the adaptive weights are misadjusted in comparison with the optimum weight solution [1]. Define the misadjustment factor η to be

$$\eta \triangleq \lim_{j \to \infty} \frac{E[\varepsilon(j)] - \varepsilon_{min}}{\varepsilon_{min}}. \qquad (4.51)$$

Using (4.3) and (4.6), it is easy to show that

$$\varepsilon(j) \triangleq \varepsilon[\mathbf{w}(j)] = \varepsilon_{min} + (\mathbf{w}(j) - \mathbf{w}_{opt})^{\dagger} \mathbf{R}(\mathbf{w}(j) - \mathbf{w}_{opt}) \qquad (4.52)$$

so that

$$\eta = \lim_{j \to \infty} \frac{E[(\mathbf{w}(j) - \mathbf{w}_{opt})^{\dagger} \mathbf{R}(\mathbf{w}(j) - \mathbf{w}_{opt})]}{\varepsilon_{min}}. \qquad (4.53)$$

To complete the above analysis, it is necessary to evaluate $Cov[\mathbf{w}(j)]$. Following [10], subtract $\mathbf{w}_{opt}$ from both sides of (4.40) to obtain

$$\mathbf{v}(j+1) = \mathbf{v}(j) - \Delta_s \hat{\nabla}_{\mathbf{w}}[\varepsilon(j)] \qquad (4.54)$$

where

$$\mathbf{v}(j) = \mathbf{w}(j) - \mathbf{w}_{opt}. \qquad (4.55)$$

Also let

$$\mathbf{g}(j) = \hat{\nabla}_{\mathbf{w}}[\varepsilon(j)] - \nabla_{\mathbf{w}}[\varepsilon(j)] \qquad (4.56)$$

represent the error in the gradient estimate, so that (4.54) becomes

$$\mathbf{v}(j+1) = \mathbf{v}(j) - \Delta_s (\nabla_{\mathbf{w}}[\varepsilon(j)] + \mathbf{g}(j)). \qquad (4.57)$$

Notice that using (4.55) in (4.5), we also have

$$\nabla_{\mathbf{w}}[\varepsilon(j)] = 2\,[\mathbf{R}\,(\mathbf{v}(j) + \mathbf{w}_{opt}) - \boldsymbol{\gamma}_{xd}]$$

$$= 2\,[(\mathbf{R}\mathbf{w}_{opt} - \boldsymbol{\gamma}_{xd}) + \mathbf{R}\mathbf{v}(j)] = 2\,\mathbf{R}\mathbf{v}(j), \quad (4.58)$$

so that the error vector in (4.57) becomes

$$\mathbf{v}(j+1) = \mathbf{v}(j) - \Delta_s\,[2\,\mathbf{R}\mathbf{v}(j) + \mathbf{g}(j)]$$

$$= (\mathbf{I} - 2\Delta_s\,\mathbf{R})\,\mathbf{v}(j) - \Delta_s\,\mathbf{g}(j)\,.$$

To simplify the analysis further, it is convenient to diagonalize and transform the coordinate frame as in (4.37). This gives

$$\tilde{\mathbf{v}}(j+1) = (\mathbf{I} - 2\Delta_s\,\Lambda)\,\tilde{\mathbf{v}}(j) - \Delta_s\,\tilde{\mathbf{g}}(j)$$

where $\tilde{\mathbf{v}}(j) = \mathbf{B}^{\dagger}\mathbf{v}(j)$ and $\tilde{\mathbf{g}}(j) = \mathbf{B}^{\dagger}\mathbf{g}(j)$. In the steady state case $(j \to \infty)$, $\tilde{\mathbf{v}}(j)$ responds only to the stationary driving function $\Delta_s\,\tilde{\mathbf{g}}(j)$ and to obtain the covariance matrix of $\mathbf{v}(j)$ in that case, form

$$\tilde{\mathbf{v}}(j+1)\tilde{\mathbf{v}}^{\dagger}(j+1) = (\mathbf{I} - 2\Delta_s\,\Lambda)\,\tilde{\mathbf{v}}(j)\,\tilde{\mathbf{v}}^{\dagger}(j)\,(\mathbf{I} - 2\Delta_s\,\Lambda) + \Delta_s^2\,\tilde{\mathbf{g}}(j)\,\tilde{\mathbf{g}}^{\dagger}(j)$$

$$- \Delta_s\,(\mathbf{I} - 2\Delta_s\,\Lambda)\,\tilde{\mathbf{v}}(j)\,\tilde{\mathbf{g}}^{\dagger}(j) - \Delta_s\,\tilde{\mathbf{g}}(j)\,\tilde{\mathbf{v}}^{\dagger}(j)\,(\mathbf{I} - 2\Delta_s\,\Lambda)\,.$$

Since $\tilde{\mathbf{v}}(j)$ is affected only by the gradient estimate of previous iterations, $\tilde{\mathbf{v}}(j)$ and $\tilde{\mathbf{g}}(j)$ are uncorrelated. This together with $E\,[\tilde{\mathbf{g}}(j)] = \mathbf{0}$ gives

$$E\,[\tilde{\mathbf{v}}(j+1)\,\tilde{\mathbf{v}}^{\dagger}(j+1)] = (\mathbf{I} - 2\Delta_s\,\Lambda)\,E\,[\tilde{\mathbf{v}}(j)\,\tilde{\mathbf{v}}^{\dagger}(j)]\,(\mathbf{I} - 2\Delta_s\,\Lambda)$$

$$+ \Delta_s^2\,E\,[\tilde{\mathbf{g}}(j)\,\tilde{\mathbf{g}}^{\dagger}(j)]\,. \quad (4.59)$$

To compute $E\,[\tilde{\mathbf{g}}(j)\,\tilde{\mathbf{g}}^{\dagger}(j)]$, we resort to (4.56) together with (4.58) and (4.43). Thus

$$\tilde{\mathbf{g}}(j) = \mathbf{B}^{\dagger}(-2e^{*}(j)\,\mathbf{x}(j) - 2\,\mathbf{R}\mathbf{v}(j))$$

$$= -2\,\mathbf{B}^{\dagger}e^{*}(j)\,\mathbf{x}(j) - 2\Lambda\tilde{\mathbf{v}}(j)$$

or equivalently

$$y(j) \stackrel{\Delta}{=} \tilde{g}(j) + 2\Lambda\tilde{v}(j) = -2B^t e^*(j)x(j).$$

Once again making use of the uncorrelated property of $\tilde{g}(j)$ and $\tilde{v}(j)$ in $E[y(j)y^t(j)]$, we get

$$E[\tilde{g}(j)\tilde{g}^t(j)] + 4\Lambda E[\tilde{v}(j)\tilde{v}^t(j)]\Lambda = 4B^t E[\,|e(j)|^2 x(j)x^t(j)]B$$

or

$$E[\tilde{g}(j)\tilde{g}^t(j)] =$$

$$4B^t E[\,|e(j)|^2 x(j)x^t(j)]B - 4\Lambda E[\tilde{v}(j)\tilde{v}^t(j)]\Lambda. \qquad (4.60)$$

With (4.60) in (4.59), it simplifies into

$$E[\tilde{v}(j+1)\tilde{v}^t(j+1)] + 2\Delta_s(\Lambda E[\tilde{v}(j)\tilde{v}^t(j)] + E[\tilde{v}(j)\tilde{v}^t(j)]\Lambda)$$

$$= E[\tilde{v}(j)\tilde{v}^t(j)] + 4\Delta_s^2 B^t E[\,|e(j)|^2 x(j)x^t(j)]B.$$

From (4.50) and (4.55), in the steady state case $E[\tilde{v}(j)] \to 0$ so that after a sufficiently large number of iterations $E[\tilde{v}(j)\tilde{v}^t(j)] = Cov[\tilde{v}(j)]$. Moreover in that case $(E[w(j)] \to w_{opt})$, from linear filter theory the error $e(j)$ is uncorrelated with the input data vector $x(j)$ and as a result, the above equation reduces into the compact form

$$\lim_{j\to\infty}\Lambda Cov[\tilde{v}(j)] + Cov[\tilde{v}(j)]\Lambda = 2\Delta_s\left(\lim_{j\to\infty}E[\,|e(j)|^2]\right)B^t R B$$

$$= 2\Delta_s\,\varepsilon_{\min}\Lambda. \qquad (4.61)$$

This in turn implies the absence of any cross-coupling in the transformed coordinate system. Thus in the steady state case $Cov[\tilde{v}(j)]$ is diagonal and hence (4.61) can be rewritten as

$$\lim_{j\to\infty}Cov[\tilde{v}(j)] = \lim_{j\to\infty}Cov[v(j)]$$

$$= \lim_{j\to\infty}Cov[w(j)] = \Delta_s\,\varepsilon_{\min}I. \qquad (4.62)$$

Thus, the weight vector does not satisfy (4.46) and even after sufficiently large number of iterations, the estimated weight vector wanders randomly around the optimum solution. As (4.62) shows, this random wandering can only be controlled by adjusting the step size Δ_s.

The misadjustment factor η defined in (4.51) and (4.53) to measure this imperfection can be simplified using (4.62). Using (4.55) in (4.53) we have

$$\eta = \lim_{j \to \infty} \frac{E[\mathbf{v}^\dagger(j)\mathbf{R}\mathbf{v}(j)]}{\varepsilon_{\min}} = \lim_{j \to \infty} \frac{E[\tilde{\mathbf{v}}^\dagger(j)\mathbf{\Lambda}\tilde{\mathbf{v}}(j)]}{\varepsilon_{\min}}$$

$$= \lim_{j \to \infty} \frac{\sum_{i=1}^{M} \lambda_i E[\,|\tilde{v}_i(j)|^2\,]}{\varepsilon_{\min}} = \Delta_s \sum_{i=1}^{M} \lambda_i = \Delta_s\, tr(\mathbf{R}).$$

Not surprisingly, Δ_s alone controls the misadjustment factor also. When the step size is decreased, the time required to attain the steady state condition increases, thereby exhibiting the trade-off between the degree of misadjustment and the adaptation speed.

Adaptive schemes based on alternate ways of estimating the unknown gradient vector have been proposed [1]. The differential steepest descent (DSD) method [10], the accelerated gradient method based on conjugate gradient descent [1] and gradient algorithms with constraints are a few examples utilizing the above approach. Of these, the DSD method obtains gradient vector estimates by a direct method that utilizes symmetric differences: Thus for a single variable w

$$\hat{\nabla}[\varepsilon(j)] = \frac{\varepsilon(w(j) + \delta) - \varepsilon(w(j) - \delta)}{2\delta}$$

which requires two distinct settings for the weight adjustment to estimate the gradient vector. Once again proceeding as before, it is possible to work out the convergence properties. Next we consider direct implementation of the Wiener solution based on the sample covariance matrix of the signal environment.

4.3.3 Direct Implementation by Inversion of the Sample Covariance Matrix

Adaptive implementation of the Wiener solution based on LMS or maximization of SNR algorithms may result in slow weight vector convergence, especially when the eigenvalue spread of the array output covariance matrix is large. In many practical applications the usefulness of the adaptive array critically depends upon the rate of convergence that can be realized. This situation occurs in radar and communication scenes where signal reception under simultaneous rejection of jamming and clutter is often essential. Since the signal environment frequently changes because of nonstationary interference signals and new jamming strategies, the adaptive processor must continuously update the weight vector to match up with the changing scenes. One possible approach to speed up convergence and avoid the convergence rate dependence on the eigenvalue distribution is to employ the Wiener solution directly, i.e., estimate the unknown covariance matrix and crosscovariance vector in (4.6) and implement them by inverting the sample covariance matrix [1, 11].

When the desired signal is absent, then $\mathbf{R} = \mathbf{R}_n$ and the optimum weight vector along the angle of incidence ω is given by

$$\mathbf{w} = \mathbf{R}_n^{-1} \mathbf{a}(\omega) , \qquad (4.63)$$

which is the same as the maximization of output SNR solution given by (4.12). The method of implementing the weight vector solutions in (4.6) and (4.63) by replacing $\mathbf{R}$, $\mathbf{R}_n$ and γ_{xd} with their respective ML estimates is known as Direct Matrix Inversion (DMI) technique.

Although the rapid convergence rate makes this method preferable to gradient based techniques, the practical difficulties associated with realizing such a solutions are by no means negligible. To start with, when N data samples are used, the formation of the sample covariance matrix requires $NM(M+1)/2$ complex multiplications. Similarly the inversion of the hermitian matrix requires another $(M^3/2+M^2)$ complex multiplications. Finally an additional (M^2+M) multiplications are required to obtain the desired weight vector. Matrix inversion can also present computational problems especially when the matrix to be inverted is ill-conditioned. Since large eigenvalue spread in the matrix to be inverted affects its ill-conditioning, the DMI approach is also susceptible to the eigenvalue spread.

The performance of the estimated weight vector based on DMI techniques can be characterized by studying the statistical properties of the associated output SNR, that has been normalized to the maximum output SNR given by (4.11). Letting ρ denote this normalized SNR, from (4.9) we have

$$\rho = \frac{(SNR)_o}{(SNR)_{max}} = \frac{|\mathbf{w}^\dagger \mathbf{a}(\omega)|^2}{\mathbf{w}^\dagger \mathbf{R}_n \mathbf{w}} \frac{1}{\mathbf{a}^\dagger(\omega) \mathbf{R}_n^{-1} \mathbf{a}(\omega)}. \tag{4.64}$$

If $\hat{\mathbf{w}}$ denotes the estimated weight vector, this gives the estimated normalized SNR to be

$$\hat{\rho} = \frac{|\hat{\mathbf{w}}^\dagger \mathbf{a}(\omega)|^2}{\hat{\mathbf{w}}^\dagger \mathbf{R}_n \hat{\mathbf{w}}} \frac{1}{\mathbf{a}^\dagger(\omega) \mathbf{R}_n^{-1} \mathbf{a}(\omega)}. \tag{4.65}$$

To simplify the analysis, first we will restrict ourselves to the case where $\boldsymbol{\gamma}_{xd}$ in (4.6) or equivalently $\mathbf{a}(\omega)$ in (4.63) is known. In that case, the sample weight vector $\hat{\mathbf{w}}_1$ corresponding to (4.6) has the form

$$\hat{\mathbf{w}}_1 = \hat{\mathbf{R}}^{-1} \boldsymbol{\gamma}_{xd} \tag{4.66}$$

and that corresponding to (4.63) takes the form

$$\hat{\mathbf{w}}_2 = \hat{\mathbf{R}}_n^{-1} \mathbf{a}(\omega). \tag{4.67}$$

Here $\hat{\mathbf{R}}$ and $\hat{\mathbf{R}}_n$ are the ML estimate (see (3.7)) of the array output covariance matrix, in presence and absence of the desired signal respectively. With (4.66) and (4.67) in (4.65), the associated normalized sample SNRs are given by

$$\hat{\rho}_1 = \frac{|\boldsymbol{\gamma}_{xd}^\dagger \hat{\mathbf{R}}^{-1} \mathbf{a}(\omega)|^2}{\boldsymbol{\gamma}_{xd}^\dagger \hat{\mathbf{R}}^{-1} \mathbf{R}_n \hat{\mathbf{R}}^{-1} \boldsymbol{\gamma}_{xd}} \frac{1}{\mathbf{a}^\dagger(\omega) \mathbf{R}_n^{-1} \mathbf{a}(\omega)}. \tag{4.68}$$

and

$$\hat{\rho}_2 = \frac{|\mathbf{a}(\omega)^\dagger \hat{\mathbf{R}}_n^{-1} \mathbf{a}(\omega)|^2}{\mathbf{a}^\dagger(\omega) \hat{\mathbf{R}}_n^{-1} \mathbf{R}_n \hat{\mathbf{R}}_n^{-1} \mathbf{a}(\omega)} \frac{1}{\mathbf{a}^\dagger(\omega) \mathbf{R}_n^{-1} \mathbf{a}(\omega)}. \tag{4.69}$$

Clearly, $\hat{\rho}_2$ has meaning only if $\hat{\mathbf{w}}_2$ is used when the desired signal is actually present and the weight adjustment $\hat{\mathbf{w}}_2$ in this case may be assumed to take place when the desired signal is absent. In case of independent, zero mean Gaussian data samples, using (3.P.8) it easily follows that

$$\hat{\rho}_2 \sim \beta(N-M+2, M-1), \qquad (4.70)$$

i.e., $\hat{\rho}_2$ is a Beta-distributed random variable with parameters $N-M+2$ and $M-1$. This result was first derived in the complex case by Reed, Mallett and Brennan [11]. Thus, the average value of $\hat{\rho}_2$ is given by

$$E[\hat{\rho}_2] = \frac{N-M+2}{N+1} \qquad (4.71)$$

and its variance is given by

$$Var(\hat{\rho}_2) = \frac{(N-M+2)(M-1)}{(N+1)^2(N+2)} \approx \frac{1}{N^2}.$$

Notice that as $N \to \infty$, $E[\hat{\rho}_2] \to 1$ and $Var(\hat{\rho}_2) \to 0$, indicating "quadratic" type of convergence for $\hat{\mathbf{w}}_2$ to the optimum solution given by (4.63). From (4.71), so long as $N \geq 2M$, we have $E[\hat{\rho}_2] \geq 1/2$, or the loss in $E[\hat{\rho}_2]$ because of nonoptimum weights is less than 3 dB. This suggests that the number of samples required to generate a useful sample covariance matrix is roughly twice the number of sensors present in the array.

Similarly, the convergence behavior of the sample weight vector, that has been estimated as in (4.66) when the signal is present, can be studied by investigating $\hat{\rho}_1$. However, instead of deriving the probability density function of $\hat{\rho}_1$ directly, it is advantageous to make use of the results for $\hat{\rho}_2$ by defining the random variable

$$\hat{\rho}_1{}' = \frac{|\hat{\mathbf{w}}_1^\dagger \mathbf{a}(\omega)|^2}{\hat{\mathbf{w}}_1^\dagger \mathbf{R} \hat{\mathbf{w}}_1} \frac{1}{\mathbf{a}^\dagger(\omega) \mathbf{R}^{-1} \mathbf{a}(\omega)}. \qquad (4.72)$$

with $\hat{\mathbf{w}}_1 = \hat{\mathbf{R}}^{-1} \gamma_{xd}$. A simple manipulation will show the relationship between $\hat{\rho}_1$ and $\hat{\rho}_1{}'$ to be

$$\hat{\rho}_1 = \frac{\hat{\rho}_1{}'}{1 + (1-\hat{\rho}_1{}')(SNR)_{max}}. \qquad (4.73)$$

Since algebraically the random variable $\hat{\rho}_1{'}$ is identical to

$$\hat{\rho}_2 = \frac{|\,\hat{\mathbf{w}}_2^\dagger\,\mathbf{a}(\omega)\,|^2}{\hat{\mathbf{w}}_2^\dagger\,\mathbf{R}_n\,\hat{\mathbf{w}}_2}\;\frac{1}{\mathbf{a}^\dagger(\omega)\,\mathbf{R}_n^{-1}\,\mathbf{a}(\omega)}\;.$$

where $\hat{\mathbf{w}}_2 = \hat{\mathbf{R}}_n^{-1}\,\mathbf{a}(\omega)$, we conclude that $\hat{\rho}_1{'}$ has the same probability density function as that of $\hat{\rho}_2$ given by (4.70). Using (4.73), it follows that

$$E[\hat{\rho}_1] = E\!\left[\frac{\hat{\rho}_1{'}}{1 + (1-\hat{\rho}_1{'})(SNR)_{max}}\right] < E[\hat{\rho}_1{'}] = E[\hat{\rho}_2]\,.$$

The above inequality implies that on the average, the output SNR achieved by using $\hat{\mathbf{w}}_1$ is less than the output SNR realized by $\hat{\mathbf{w}}_2$. This behavior may be explained by noticing that, the presence of the desired signal increases the time required to obtain accurate estimates of $\mathbf{R}$, compared to the time required for accurate estimates of $\mathbf{R}_n$ when the desired signal is absent. From (4.73), it also follows that for weak signals $((SNR)_{max} < 1)$ the difference in performance obtained using $\hat{\mathbf{w}}_1$ and $\hat{\mathbf{w}}_2$ is negligible.

Since $\hat{\mathbf{w}}_2$ outperforms $\hat{\mathbf{w}}_1$ when $(SNR)_{max} > 1$, it is advantageous to use $\hat{\mathbf{w}}_2$ or estimate $\hat{\mathbf{R}}$ after removing the desired signal component $\mathbf{u}(t) = d(t)\,\mathbf{a}(\omega)$ in (4.8) from the received vector $\mathbf{x}(t)$. This allows to generate $\hat{\mathbf{R}}_n$ from desired signal free noise vectors. However the method of eliminating the desired signal component through the use of

$$\mathbf{S} = \hat{\mathbf{R}} - P\,\mathbf{a}^\dagger(\omega)\,\mathbf{a}(\omega)$$

does not lead to any improvement over $\hat{\mathbf{w}}_1$ [1]. This can be verified by noticing that the weight vector generated by $\mathbf{S}$ can be rewritten as

$$\hat{\mathbf{w}}_3 = \mathbf{S}^{-1}\boldsymbol{\gamma}_{xd} = [\hat{\mathbf{R}} - P\,\mathbf{a}^\dagger(\omega)\,\mathbf{a}(\omega)]^{-1}\boldsymbol{\gamma}_{xd}$$

$$= \left(\frac{1}{1 - P\,\mathbf{a}^\dagger(\omega)\,\hat{\mathbf{R}}^{-1}\,\mathbf{a}(\omega)}\right)\hat{\mathbf{R}}^{-1}\boldsymbol{\gamma}_{xd} = k\,\hat{\mathbf{w}}_1\,.$$

Since $\hat{\mathbf{w}}_3$ is a scalar multiple of $\hat{\mathbf{w}}_1$, its convergence properties are the same as those of $\hat{\mathbf{w}}_1$, and hence no transient performance improvement can be realized in this manner. For such an improvement in performance $\hat{\mathbf{w}}_3$ must be a scalar multiple of $\hat{\mathbf{w}}_2$, and this can be achieved only by "cleaning" the data prior to estimating the array output covariance matrix.

Next, we remove the assumption that γ_{xd} is known in (4.66) and consider the more realistic case where the array output covariance matrix $\mathbf{R}$ and the crosscovariance vector γ_{xd} are unknown. In this case the estimated weight vector corresponding to the Wiener solution in (4.6) can be written as

$$\hat{\mathbf{w}}_4 = \hat{\mathbf{R}}^{-1} \hat{\gamma}_{xd} \tag{4.74}$$

where from (4.4)

$$\hat{\gamma}_{xd} = \frac{1}{N} \sum_{n=1}^{N} \mathbf{x}(n) d^*(n) \tag{4.75}$$

represents the ML estimate for γ_{xd}. Once again the transient response characteristics can be studied by investigating the normalized sample SNR

$$\hat{\rho}_4 = \frac{|\hat{\mathbf{w}}_4^\dagger \mathbf{a}(\omega)|^2}{\hat{\mathbf{w}}_4^\dagger \mathbf{R}_n \hat{\mathbf{w}}_4} \frac{1}{\mathbf{a}^\dagger(\omega) \mathbf{R}_n^{-1} \mathbf{a}(\omega)}. \tag{4.76}$$

It is difficult to evaluate the probability density function of (4.76) in a closed form, and alternatively the mean square error may be taken to evaluate the transient response. Since $\mathbf{w}_{opt}$ results in the irreducible global minimum error ε_{min}, the convergence properties of the sample mean square error associated with $\hat{\mathbf{w}}_4$ is a direct indication of those of $\hat{\mathbf{w}}_4$ itself. This is even more meaningful in Coherent SideLobe Cancellation (SCLC) systems, where a high gain antenna points in the direction of the desired signal (Fig. 4.4). The auxiliary antennas are designed so that their gain patterns approximate the average sidelobe level of the main antenna gain pattern. As a result, the auxiliary antennas are able to provide replicas of jamming signals appearing in the sidelobes of the main antenna pattern, which are cancelled out, thereby providing an interference free array output signal. Quick

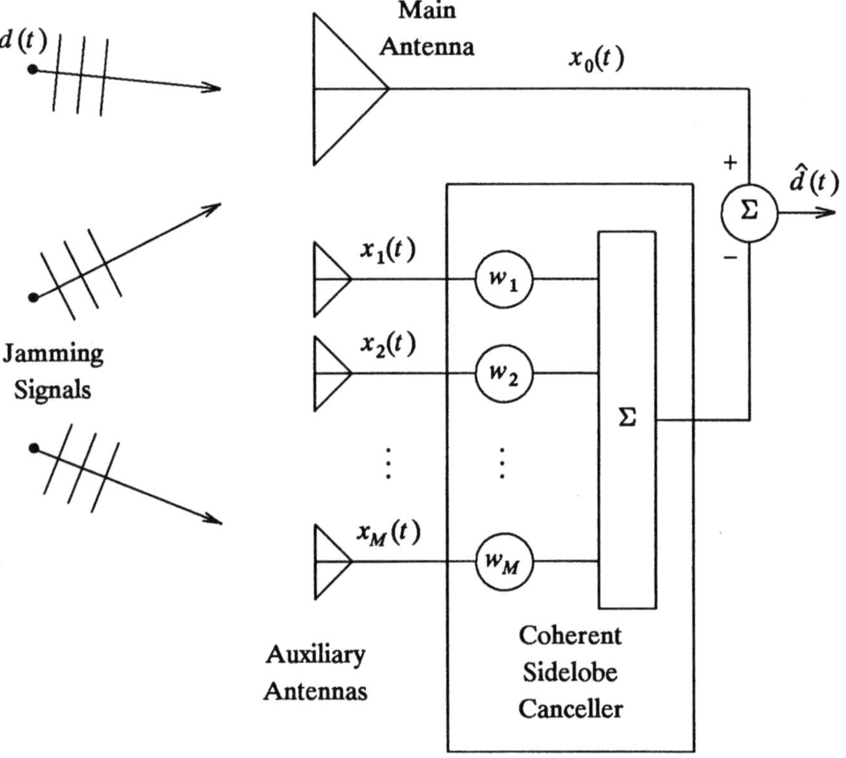

Fig. 4.4 Coherent sidelobe cancellation (CSLC) system.

adaptation to the optimum weight vector solution in this case will minimize the output residue power and hence minimize the interference signal component at the array output. The configuration in Fig. 4.4 can also stand for the situation represented by (4.2) provided $x_0(t) = d(t)$. In that case the error signal is given by

$$e(n) = d(n) - \hat{\mathbf{w}}_4^{\dagger}\mathbf{x}(n) \qquad (4.77)$$

and the estimated mean square error can be expressed as

$$\hat{\varepsilon}(\hat{\mathbf{w}}_4) = \frac{1}{N}\sum_{n=1}^{N}|e(n)|^2 = \hat{P} - \hat{\mathbf{w}}_4^{\dagger}\hat{\gamma}_{xd} - \hat{\gamma}_{xd}^{\dagger}\hat{\mathbf{w}}_4 + \hat{\mathbf{w}}_4^{\dagger}\hat{\mathbf{R}}\hat{\mathbf{w}}_4$$

$$= \hat{P} - \hat{\gamma}_{xd}^{\dagger}\hat{\mathbf{R}}^{-1}\hat{\gamma}_{xd} \qquad (4.78)$$

where $\hat{\mathbf{R}}$, $\hat{\mathbf{w}}_4$, $\hat{\gamma}_{xd}$ are as defined in (3.7), (4.74) – (4.75) respectively and

$$\hat{P} = \frac{1}{N} \sum_{n=1}^{N} |d(n)|^2 \tag{4.79}$$

represents the sample power estimator of the desired signal. The transient behavior can be studied by evaluating the statistical properties of $\hat{\varepsilon}(\hat{\mathbf{w}}_4)$. Towards this, if the augmented $(M+1) \times 1$ vector $[\mathbf{x}(n), d(n)]^T$ is assumed to be zero mean jointly Gaussian with covariance matrix

$$\begin{bmatrix} \mathbf{R} & \gamma_{xd} \\ \gamma_{xd}^\dagger & P \end{bmatrix}$$

then from (3.A.8), the 1×1 lower right hand corner Schur complement

$$N(\hat{P} - \hat{\gamma}_{xd}^\dagger \hat{\mathbf{R}}^{-1} \hat{\gamma}_{xd}) = N \hat{\varepsilon}(\hat{\mathbf{w}}_4)$$

$$\sim CW(N-M, 1, P - \gamma_{xd}^\dagger \mathbf{R}^{-1} \gamma_{xd} = \varepsilon_{\min}) \tag{4.80}$$

or using (3.A.20), we have (see also (3.18))

$$\frac{N \hat{\varepsilon}(\hat{\mathbf{w}}_4)}{\varepsilon_{\min}} \sim \chi^2(N-M). \tag{4.81}$$

This gives the mean and variance of the sample mean square estimator $\hat{\varepsilon}$ in (4.78) to be

$$E[\hat{\varepsilon}(\hat{\mathbf{w}}_4)] = \left(1 - \frac{M}{N}\right) \varepsilon_{\min} \tag{4.82}$$

$$Var[\hat{\varepsilon}(\hat{\mathbf{w}}_4)] = \frac{1}{N}\left(1 - \frac{M}{N}\right) \varepsilon_{\min}^2. \tag{4.83}$$

Following (4.51) – (4.53), the difference between the actual MSE $\varepsilon(\hat{\mathbf{w}}_4)$ and the minimum MSE $\varepsilon_{\min}$ also is a direct measure of the quality of the array performance relative to the optimum and this gives the sample misadjustment factor $\hat{\eta}$ to be

$$\hat{\eta} = \frac{\varepsilon(\hat{\mathbf{w}}_4) - \varepsilon_{min}}{\varepsilon_{min}} = \frac{\Delta\mathbf{w}^\dagger \mathbf{R} \Delta\mathbf{w}}{\varepsilon_{min}} \qquad (4.84)$$

where

$$\Delta\mathbf{w} = \hat{\mathbf{w}}_4 - \mathbf{w}_{opt} = \hat{\mathbf{R}}^{-1}\hat{\gamma}_{xd} - \mathbf{R}^{-1}\gamma_{xd} \,.$$

The conditional normality of the regression coefficient $\hat{\mathbf{w}}_4$ given $\hat{\mathbf{R}}$ (see (3.A.9)) together with (3.A.10) can be used to determine the probability density function of $\hat{\eta}$. Letting

$$\hat{\eta} = r^2 , \qquad (4.85)$$

the probability density function of r can be shown to be [12]

$$f_r(r) = 2\frac{N!}{(N-M)!\,(M-1)!}\frac{r^{2M-1}}{(1+r^2)^{N+1}} , \quad 0 < r < \infty , \quad (4.86)$$

i.e., the random variable r has an F-distribution and hence

$$b = \frac{1}{1+r^2}$$

is a Beta-distributed random variable with parameters $N-M+1$ and M. Thus

$$E[r^2] = E[\hat{\eta}] = \frac{N}{N-M} \qquad (4.87)$$

and

$$Var(r^2) = Var(\hat{\eta}) = \frac{NM}{(N-M)^2(N-M-1)} . \qquad (4.88)$$

Moreover, from (3.A.8) – (3.A.10) it also follows that $\hat{\varepsilon}(\hat{\mathbf{w}}_4)$ in (4.78) is statistically independent of $\hat{\mathbf{w}}_4$ and $\hat{\mathbf{R}}$.

From (4.88), the output residue power is within 3 dB of the optimum value only after using $2M$ distinct data samples and simultaneously the variance of this residue power approaches zero quadratically, thereby indicating rapid convergence for the DMI technique independent of the signal environment. A comparison of (4.66) and (4.74) will also indicate that the transient performance of $\hat{\mathbf{w}}_1$ cannot be inferior to that of $\hat{\mathbf{w}}_4$.

The transient response obtained using $\hat{\mathbf{w}}_1$ and $\hat{\mathbf{w}}_4$ has also been found to be superior to that obtained from the LMS algorithm, which has an iteration period equal to the intersample period of the DMI algorithm [1]. However, the transient response of the LMS algorithm depends on the step size, which can be made arbitrarily small, thereby requiring an arbitrary large number of iterations for its convergence. The LMS response rate also depends on the initial weight vector setting. A good initial setting can result in excellent transient response characteristics. For example, by setting $\mathbf{w}(0) = \gamma_{xd}$ the initial LMS algorithm response can be greatly improved, since this initial starting condition biases the array toward the desired signal direction and provides a high output SNR. Although the weight vectors exhibit convergence $(Var(\hat{p}) \rightarrow 0)$ in case of DMI algorithms, the weight vector variance remains constant for the LMS algorithm and depends only on the step size (see (4.62)). For these reasons, a comparison of the DMI and the LMS algorithms is not entirely satisfactory.

Although, unlike the LMS algorithm, the convergence speed of the DMI algorithms is insensitive to the eigenvalue spread in $\mathbf{R}$, this eigenvalue spread can generate ill-conditioned matrices that cannot be inverted with sufficient accuracy. This can be controlled by increasing the number of bits available to perform the matrix inversion. So long as the sample covariance matrix has an eigenvalue spread that is less than a critical value, which depends on the number of available bits, the DMI algorithms are insensitive to eigenvalue spread.

Problems

1. Let $\mathbf{x}$ and $\mathbf{y}$ be m and n dimensional jointly Gaussian random vectors with mean values and covariances given by

$$E[\mathbf{x}] = \mu_x \, , \quad E[\mathbf{y}] = \mu_y \, ,$$

$$Cov\,[\mathbf{x}] \triangleq E\,[(\mathbf{x}-\boldsymbol{\mu}_x)(\mathbf{x}-\boldsymbol{\mu}_x)^\dagger] = \mathbf{R}_x\,,$$

$$Cov\,[\mathbf{y}] = \mathbf{R}_y$$

and

$$Cov\!\begin{pmatrix}\mathbf{x}\\\mathbf{y}\end{pmatrix} = \begin{bmatrix} \mathbf{R}_x & \mathbf{R}_{xy} \\ \mathbf{R}_{xy}^\dagger & \mathbf{R}_y \end{bmatrix} \triangleq \mathbf{R}\,.$$

Then

$$f_{\mathbf{x}}(\mathbf{x}) = \frac{1}{|\pi\mathbf{R}_x|}\, e^{-(\mathbf{x}-\boldsymbol{\mu}_x)^\dagger \mathbf{R}_x^{-1}(\mathbf{x}-\boldsymbol{\mu}_x)}\,.$$

Show that the conditional probability density function $f_{\mathbf{x}|\mathbf{y}}(\mathbf{x}\,|\,\mathbf{y})$ of $\mathbf{x}$ given $\mathbf{y}$ is also Gaussian with

$$f_{\mathbf{x}|\mathbf{y}}(\mathbf{x}\,|\,\mathbf{y}) = \frac{1}{|\pi\mathbf{D}|}\, e^{-(\mathbf{x}-\mathbf{a})^\dagger \mathbf{D}^{-1}(\mathbf{x}-\mathbf{a})}$$

where

$$\mathbf{a} = \boldsymbol{\mu}_x + \mathbf{R}_{xy}\,\mathbf{R}_y^{-1}(\mathbf{y}-\boldsymbol{\mu}_y)$$

and

$$\mathbf{D} = \mathbf{R}_x - \mathbf{R}_{xy}\,\mathbf{R}_y^{-1}\mathbf{R}_{xy}^\dagger\,.$$

2.   If the desired signal is uncorrelated with the rest of the signals and noise present in the scene, show that the Wiener solution (4.6) reduces to

$$\mathbf{w}_{opt} = \alpha\,\mathbf{R}_n^{-1}\,\mathbf{a}(\omega)$$

where $\alpha = P\,/\,(1+P\,\mathbf{a}^\dagger(\omega)\,\mathbf{R}_n^{-1}\,\mathbf{a}(\omega))$ and $\mathbf{a}(\omega)$ represents the direction vector associated with the desired signal.

3.   **The minimum noise variance performance measure.** For the received signal $\mathbf{x}(t)$ in (4.8) in an uncorrelated noise scene, let $y_i(t)$ represent the aligned signal at the ith sensor, i.e.,

$$y_i(t) = d(t) + \sqrt{M}\, e^{j\,d_i\,\omega}\,n_i(t)\,, \quad i = 1, 2, \cdots, M\,.$$

The array output $z(t) = \mathbf{w}^\dagger \mathbf{y}(t)$, under the constraint that the sum of the array weights is unity, takes the form

$$z(t) = d(t) + \sum_{i=1}^{M} w_i^* n_i{}'(t), \quad n_i{}'(t) = \sqrt{M}\, e^{j\, d_i\, \omega}\, n_i(t).$$

Show that the weight vector that minimizes the variance of $z(t)$ is given by

$$\mathbf{w}_{MV} = \frac{\mathbf{R}_n^{-1} \mathbf{1}}{\mathbf{1}^\dagger \mathbf{R}_n^{-1} \mathbf{1}}$$

where $\mathbf{1} = \begin{bmatrix} 1, 1, \cdots, 1 \end{bmatrix}^T$.

4. **Multiple signal estimation.** Consider the received signal model

$$\mathbf{x}(t) = \mathbf{A}\,\mathbf{u}(t) + \mathbf{n}(t) \tag{4.P.1}$$

where $\mathbf{A}$ $(M \times K)$ is known and the signal vector $\mathbf{u}(t)$ $(K \times 1)$ may be random or unknown. The additive noise is assumed to be zero mean with covariance matrix $\mathbf{R}_n$.

a) Assume $\mathbf{u}(t)$ and $\mathbf{n}(t)$ to be independent Gaussian processes with $E[\mathbf{u}(t)] = \boldsymbol{\mu}$ and $Cov[\mathbf{u}(t)] \triangleq E[(\mathbf{u}(t)-\boldsymbol{\mu})(\mathbf{u}(t)-\boldsymbol{\mu})^\dagger] = \mathbf{R}_u$. Show that the MMSE criterion leads to the estimator

$$\hat{\mathbf{u}}(t) = \boldsymbol{\mu} + \mathbf{R}_u \mathbf{A}^\dagger (\mathbf{A}\mathbf{R}_u \mathbf{A}^\dagger + \mathbf{R}_n)^{-1}(\mathbf{x}(t) - \mathbf{A}\boldsymbol{\mu}). \tag{4.P.2}$$

(Hint : use the conditional mean.)

b) The desired signal vector $\mathbf{u}(t)$ is non-Gaussian random signal. Show that the best linear estimate of the form

$$\hat{\mathbf{u}}(t) = \mathbf{a} + \mathbf{B}\mathbf{x}(t) \tag{4.P.3}$$

that minimizes the MSE given by $E[\,|\mathbf{u}(t) - \hat{\mathbf{u}}(t)|^2\,]$ is obtained for the choice

$$\mathbf{a} = (\mathbf{I} - \mathbf{B}\mathbf{A})\boldsymbol{\mu} \quad \text{and} \quad \mathbf{B} = \mathbf{R}_n \mathbf{A}^\dagger (\mathbf{A}\mathbf{R}_u \mathbf{A}^\dagger + \mathbf{R}_n)^{-1}.$$

c)   The desired signal vector $\mathbf{u}(t)$ is nonrandom but unknown and the noise is Gaussian. Show that the best MSE is obtained for

$$\hat{\mathbf{u}}(t) = (\mathbf{A}^{\dagger}\mathbf{R}_n^{-1}\mathbf{A})^{-1}\mathbf{A}^{\dagger}\mathbf{R}_n^{-1}\mathbf{x}(t).$$

5.   **Adaptive algorithm based on maximization of SNR [7].**

a)   With

$$\alpha^2(j) \triangleq \frac{|\mathbf{w}^{\dagger}(j)\mathbf{a}(\omega)|^2}{\mathbf{w}^{\dagger}(j)\mathbf{R}_n\mathbf{w}(j)}$$

representing the instantaneous SNR of the desired signal with direction vector $\mathbf{a}(\omega)$, show that maximization of SNR transforms the adaptive equation

$$\mathbf{w}(j+1) = \mathbf{w}(j) + \Delta_s \nabla[\alpha^2(j)]$$

into

$$\mathbf{w}(j+1) = \mathbf{w}(j) + 2\Delta_s \mu(j)\left[\mathbf{a}(\omega) - \mu^*(j)\mathbf{R}_n\mathbf{w}(j)\right] \quad (4.P.4)$$

where

$$\mu(j) = \frac{\mathbf{w}^{\dagger}(j)\mathbf{a}(\omega)}{\mathbf{w}^{\dagger}(j)\mathbf{R}_n\mathbf{w}(j)}.$$

b)   Suppose $\mathbf{w}(0)$ is selected such that $\mathbf{w}(j) \to \mathbf{w}_{SNR}$ given by (4.12). Show that (4.P.4) simplifies into the linear equation

$$\mathbf{w}(j+1) = \mathbf{w}(j) + 2\Delta_s \alpha_0\left[\mathbf{a}(\omega) - \alpha_0^*\mathbf{R}_n\mathbf{w}(j)\right] \quad (4.P.5)$$

where $\alpha_0 = \lim_{j \to \infty} \mu(j)$.

c)   Establish mean square convergence for (4.P.5).

6.   Derive the relationship given by (4.73).

7.   Show that [12]

$$E[\hat{\rho}_1] = \left(\frac{N-M+2}{N+1}\right)\left[1 + \sum_{i=0}^{\infty}[-(SNR)_{max}]^i \prod_{j=1}^{i}\left(\frac{j+M-2}{N+j+1}\right)\right].$$

This gives the dependence of $E[\hat{\rho}_1]$ on the maximum output SNR.

8. Show that the random variable r in (4.85) has an F-distribution.

References

[1] R. A. Monzingo and T. W. Miller, *Introduction to Adaptive Array*. New York: John Wiley and Sons, 1980.

[2] R. T. Compton, *Adaptive Antennas*. Englewood Cliffs, NJ: Prentice Hall, 1988.

[3] B. Widrow, P. E. Mantey, L. J. Griffiths, and B. B. Goode, "Adaptive antenna systems," *Proc. IEEE*, vol. 55, pp. 2143-2159, Dec. 1967.

[4] P. E. Mantey and L. J. Griffiths, "Iterative least square algorithms for signal extraction," *Proc. of the 2nd Hawaii Conf. on System Sciences*, pp. 767-770, 1969.

[5] E. B. Manoukian, *Modern Concepts and Theorems of Mathematical Statistics*. New York: Springer-Verlag, 1985.

[6] H. L. Van Trees, *Detection, Estimation, and Modulation Theory, Part I*. New York: John Wiley and Sons, 1968.

[7] L. E. Brennan and I. S. Reed, "Theory of adaptive radar," *IEEE Trans. Aerosp. Electron. Syst.*, vol. AES-9, pp. 237-252, Mar. 1973.

[8] B. D. Van Veen and K. M. Buckley, "Beamforming: A versatile approach to spatial filtering," *IEEE ASSP Magazine*, vol. 5, pp. 4-24, Apr. 1988.

[9] L. J. Griffiths, "Signal extraction using real-time adaptation of a linear multichannel filter," Ph.D. Dissertation, Stanford Univ., Dec. 1967.

[10] B. Widrow and J. M. McCool, "A comparison of adaptive algorithms based on the methods of steepest descent and random search," *IEEE Trans. Antennas Propag.*, vol. AP-24, pp. 615-638, Sept. 1976.

[11] I. S. Reed, J. D. Mallett, and L. E. Brennan, "Rapid convergence rate in adaptive arrays," *IEEE Trans. Aerosp. Electron. Syst.*, vol. AES-10, pp. 853-863, Nov. 1974.

[12] T. W. Miller, "The transient response of adaptive arrays in TDMA systems," Ph.D. Dissertation, The Ohio State Univ., 1976.

Index